The Visual World of Shadows

The Visual World of Shadows

Roberto Casati and Patrick Cavanagh

The MIT Press
Cambridge, Massachusetts
London, England

This book was set in Stone by Toppan Best-set Premedia Limited. Printed and bound in the United States of America.

Library of Congress Cataloging-in-Publication Data

Names: Casati, Roberto, 1961– author. | Cavanagh, Patrick, author.
Title: The visual world of shadows / Roberto Casati and Patrick Cavanagh.
Description: Cambridge, MA : MIT Press, [2019] | Includes bibliographical
    references and index.
Identifiers: LCCN 2018029631 | ISBN 9780262039581 (hardcover : alk. paper)
Subjects: LCSH: Visualization. | Shades and shadows.
Classification: LCC BF241 .C367 2019 | DDC 152.14—dc23 LC record available at
    https://lccn.loc.gov/2018029631

10 9 8 7 6 5 4 3 2 1

Contents

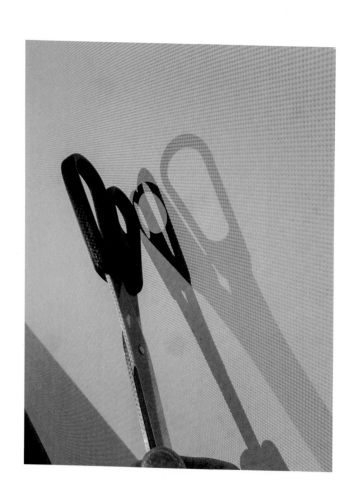

Preface

This book is about the way our visual system reads, understands, and uses shadows. The main tenet of the book is that shadows cut deep in the structure of the brain. Their peculiar nature makes them the target of extremely specific computations, which at times enhance the shadows, at times edit them out, and at times use them for various purposes, in particular for retrieving the three-dimensional structure of the environment. Studying shadow perception reveals a vast array of sophisticated, hidden mechanisms of the visual brain.

Shadows are a ubiquitous feature of our visual environment, in which so many objects block the light that is our principal source of information. Given their perturbation of light, and their high contrast, shadows are a massive problem for vision which must distinguish them from full-blown objects or material features of objects. Despite their nuisance, shadows are also an invaluable resource as their dance with light and their visibility can signal objects' presence, location, shape, and size, among other properties. We know that much of the information from shadows is processed automatically and then may be accessed consciously, on demand, as it were. What is intriguing is that once used, shadows end up in the visual dustbin—they are hardly noticed or remembered—from where they can be retrieved only in certain special circumstances. Try to recall the shadows in the room you just left. Now go back there and look at all of them; you will see that most of them went unnoticed.

Because of the realism they offer and because of their visual and conceptual complexity, shadows have long fascinated visual artists: painters, graphic designers, movie makers. Their explorations—images, paintings, sketches on cave walls—constitute a rich vein of visual experiments offering forty thousand years of documented discoveries. Each painting is in essence an experiment of how to arrange pigments on a surface to produce impressions of light, shadow, surface, and depth, properties that the pigments themselves lack but which are nevertheless triggered in our brains by the arrangements that artists have discovered. Our work can be seen as a way to pay tribute to this incredibly sophisticated collective visual investigation. Painters have struggled with the difficulties of depicting shadows, so much so that shadows—after a brief, spectacular showcase in ancient Roman paintings and mosaics—are almost absent from pictorial art up to the Renaissance and then are hardly present outside traditional Western art.

Shadows are extremely useful ingredients of visual knowledge. *Chiaroscuro* (shading) provides precious information about the three-dimensional structure of objects, about their shape. Cast shadows help to signal the distance between an object and the surface on which its shadow falls. Humans have known this for a long time, before vision scientists ever attempted to systematize shadow perception. In this book, we shall see a number of examples in which shadows provide subtle information about the spatial structure of the world, in "automatic" and unconscious ways through the mechanisms of visual perception, as opposed to the "reflective" and conscious ways of cognition. In their struggles with shadows, painters have uncovered some of the basic principles of visual compression: simplifications of the scene that are nevertheless convincing to the eye. Astronomers have also used shadows to very different ends: to reach out into an otherwise inaccessible space and measure it. Johannes Kepler, the king of astronomy of the early seventeenth century, went so far as to say that "all astronomic discoveries proceed through light and shadow." Galileo provided a dramatic illustration of this claim: his telescopic sighting of the phases of Venus struck the fatal blow to the Ptolemaic system.

The field covered by the book can be summarized in the diagram shown on the next page. Here we see that different disciplines study different aspects of shadows. Two boxes intersect: the *reality box*, which comprises the informational properties physically present in shadows; and the *representation box*, which includes all the brain does with its representations of shadows. Something lies outside the intersection. *Some* informational properties of shadows are not used by the brain; conversely, the brain is very happy with *some* representations of impossible shadows, that is, of shadows that cannot occur in nature. These outliers are interesting for two reasons. On the one hand, even if the brain does not use certain *regularities* of the information in shadows, a machine may do so in its stead. Thus it is important to characterize the information in shadow, no matter what use biological vision makes of it. On the other hand, the many shadow illusions that are generated by the discrepancy between shadow reality and shadow representation tell us quite a lot about the functioning of the brain. The visual brain seems to be rather choosy in its endeavors, a situation we can capture by laying out the *properties* that determine what is a shadow and what is not, whose shadow it is, whether it is a good or a bad shadow for that object, and how a certain shadow can lend a hand in making scene shapes and distances visible.

Chapters 1 and 2 draw an inventory of information retrievable from shadows—be it used or not by the visual system. Here we cover some classical "properties from shadow" examples: shape from shadow, size from shadow, distance, but also other, less-traveled paths. It is amazing to see how much lies hidden in shadow, ripe for us to discover.

The language of these first two chapters is a language of *laws and regularities*. But is the visual system completely attuned to them, able to detect them? Does it use this bewildering wealth of information? It turns out that things are not so straightforward. Thus chapter 3 reviews what science knows about the relevant aspects of the visual brain, and investigates in particular the early and middle stages of vision, distinguishing between

Do boats fly? The separation of the shadow from the boat indicates that the boat is floating above the surface on which the shadow is cast, discussed in chapter 1. Image credit: Danos Kounenis.

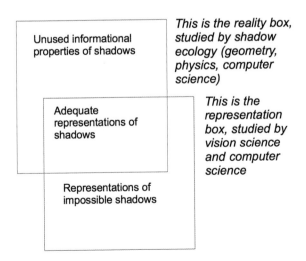

The field covered in this book.

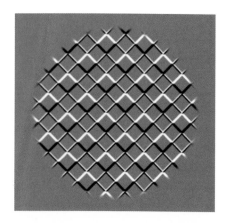

Turn the page upside down, wait a couple of seconds, and see the bumps turn into hollows—why does this happen? We discuss this phenomenon in chapter 4. Image credit: Pascal Mamassian.

Does the tree own the shadow behind it? The connection is missing. Image credit: PC, Toronto, October 2017.

measurement and *inference*. Shadows are visual entities and must be separated from other objects in the scene. Edges are key elements in the retrieval of shadows. The chapter deals in particular with the distinction between edges of illumination and edges of reflectance, that is, edges that signify an accident of illumination and edges that correspond to stable features of the environment. This discussion sets out the path for the following chapters, whose language is no longer that of laws and regularities but that of *computational constraints* imposed by shadow properties.

This path has three main steps: *labeling*, *ownership*, and *mission*.

The shadow *mission* forms the core of this research; all other steps follow in service of this goal. Shadows do inform us about shape, position, size, and many other aspects detailed in chapters 1 and 2; but can the brain really take advantage of these informational riches? If it does, how does it do so? This is the topic of chapter 4. It shows the surprising subtleties of the work performed by the visual brain when computing some aspect of shadows' contribution to the reconstruction of a three-dimensional scene. Chapter 4 also notes that the end of the shadow mission, the discounting of shadows, is of critical importance. Once their information is extracted, shadow borders are removed from further scene analysis, and their shapes are demoted in salience: they attract little or no attention. This is as it should be for nonobject, accidental effects of illumination. Although vision pays little attention to them once they have served their purpose, here we try to bring to shadows the recognition they deserve for their contributions.

The information a cast shadow carries about an object cannot be put to any use unless we can trace the shadow back to that very object. This is the problem of *shadow ownership*, the topic of chapter 5. It would be nice if each shadow came with a leash tying it to the object that casts it; sure enough, in many situations, objects are "glued" to their shadows, as we all are when we stand in the sun, stepping on our own shadow, but this is not the general case. Here, too, some interesting ownership errors—cases of shadow *capture* or *repulsion*—demonstrate the existence of specific mental rules.

But ownership is not the end of the story. Perception does not pointlessly bind together any pair of objects. To tie a shadow to its owner, vision must first *label* shadows, that is, determine that certain dark zones of the visual fields are not paint on a surface or smoke in the air but rather accidents of illumination. Accordingly, in chapter 6 we discuss the many ways one can mistake a shadow for paint or an object and vice versa—a fact that painters both discovered and treasured. Labeling a shadow is not yet consciously perceiving a shadow as a shadow, and we will explore the distinction between unconscious shadow labeling and conscious appreciation of shadow character. As visual disruption offers a good way to find out how the visual system works, we discuss various conditions that disrupt both shadow labeling and character.

Thus we know that the visual system is able to tell shadows from nonshadows, and once it solves ownership, it is ready to tap into the informational riches of shadows we discuss in chapters 1 and 2: only after labeling and ownership are achieved is everything in place for shadows to accomplish their *mission* that we describe in chapter 4.

We should step back and see the situation from the ecological and evolutionary point of view. Cast shadows are ubiquitous and peculiar ingredients of the visual scene. They are all of the same kind, no matter the object they are shadow of, and in spite of carrying information about the objects that cast them, in the end they are dark spots distributed here and there throughout the scene. When vision parses the scene, it cannot ignore them. Imagine an analogy with linguistic parsing: when listening to speech, here and there a repetition is produced, an echo, with some systematicity. We might look for a way to make sense of the echoes, to find them an informational role, a meaning. In any event, they are hard to ignore.

A brain that evolved in a shadow-rich environment had a strong incentive to solve what we might call the shadow-labeling problem: this would help a lot in the management of visual noise. The solution to the problem of shadow ownership, on the other hand, would be of marginal significance unless shadows were used for something. Why care who owns what shadow? Thus shadow ownership seems to be controlled by the uses we can put shadows to. Now, given the diverse types of information involved, and given the complex geometric transformations that take any complex three-dimensional object to its shadow on a surface that itself is geometrically complex, under many different light conditions at different times of the day and on different days of the year, we may have little hope that a genius geometrical brain emerged from natural selection, a kind of internal Euclid-Descartes-Desargues-Poncelet-Klein homunculus able to squeeze out the forbidding trigonometry of the shadow-rich visual scene. What we can reasonably expect is the discovery and deployment, by evolution, of

The vertical stripes appear to be paint of different colors, but they are actually shading effects on a uniformly colored metal surface. Why is the visual system fooled by these stripes and not by the shadow under the bar? Image credit: RC, Paris, April 2017.

Shadows are noisy. Image credit: RC, Barbizon, May 2017.

Euclid's and René Descartes's contributions to geometry are universally acknowledged. Gérard Desargues (1591–1661) produced the first working theory of shadow projections, and Jean-Victor Poncelet (1788–1867) and Felix Klein (1849–1925) formalized the general theory of projections. One way or another, all these authors used or could have used shadows in their demonstrations.

rough and robust heuristics that hold in a large enough number of typical lighting circumstances. These heuristics may build on each other or "talk to" each other but may also be mutually independent—a battery of internal x-from-shadow-computing machines or homunculi, each with its own well-specified assignment. And this is precisely what research ends up demonstrating.

The laws that ought to govern cognitive processing—geometry and consideration of physical constraints—do not quite match the poorly understood heuristics that are used rapidly and unconsciously by the visual system. Indeed, the business of vision science is precisely to find these hidden rules governing shadow perception. Sometimes these rules are quite surprising indeed: vision accepts impossible shadows and rejects quite possible ones.

Let's move on from here. You now sit in your living room armchair. Do you remember the shadows cast by the lamp on your desk? Probably not—although they were available to you countless of times in the past. Shadow representations require extra effort to achieve a place in memory. It looks as if a powerful filter unclogs the visual scene of its shadows. But, of course, it is not as if you saw reality like a shadowless cartoon. The final perceptual movie still has shadows in it; they are just carrying a kind of "do not pay attention to me" label that blunts their salience.

Chapter 7 explores the interactions between perception and what each of us knows, or believes we know, about shadows. If the visual system decides that certain areas should be labeled as shadows (chap. 6, once more), what other areas of the brain know about that, and is the knowledge stored here consistent with knowledge stored elsewhere? The cognitive conceptual system oversees the categorization of objects, allows us to make complex inferences about their properties, and provides the rich web of concepts that supports linguistic communication and practical decision making and action. To be well informed and effective, this system talks to many other systems, such as vision, and because of the way we are hardwired, it accepts vision's description of the scene in front of us, with occasional complaints about misjudgments that in the end have little influence on what we see (think of illusions). It is by looking at these various interfaces that we can probe the particular views the conceptual system has on shadows. Is that over there really a shadow? To what extent can we use shadows to guide actions? Can we control our own shadow? What can we infer from the fact that we know that a shadow appears in a certain spot? Which metaphors are supported by shadow talk?

The mental representation of shadows appears to be a patchwork of disparate, incomplete, and at times mutually contradictory elements. Under certain conditions, we may even doubt whether a certain colored area is a shadow or light. In hindsight, it may be interesting to speculate about this uncertainty. Perhaps the low salience of shadows, or their extremely focused mission, is not enough to render them interesting to systematic thought. Perhaps shadows have too many strange or surprising properties. Perhaps the fact that shadows are filtered out in perception makes us conceptually (partially) blind to them.

Spurious "shadow" under the table generated by absence of snow—an analogical snow shadow, discussed in chapter 7. Image credit: RC, Etna, New Hampshire, 2014.

Chapters 8 and 9 use the theoretical apparatus that we have developed for shadows to understand some neighboring perceptual phenomena: illumination (the "negative" of shadow casting), reflections (akin to shadows, these are "images" of objects that increase the visual population with items that are not full-blown, concrete objects and still compete with objects), and transparency (again akin to shadows, these are "layers" of visual entities). First, we study what is the *information* available and, second, how it is used to understand the scene (its *mission*). Then we ask if there is a process to establish *ownership*: what object is paired with what light spot or reflection? Then, what is it to be a reflection, or a mirror, or a transparent overlay? Finally, what are the properties required to *label* a surface a mirror or a transparency, and are there any physical properties that can be ignored? The analogies point to possible commonalities of visual processing in all these cases.

Chapter 10 capitalizes on the previous discussion and analyzes the complexities of shadow representation in art history. This is a "last but not least" chapter. You may expect that after having shown how shadow perception works, we would just enhance the stimuli you would typically find in textbooks and scientific articles by adding a few examples from art. Rather, we propose to consider artists as fellow cognitive scientists: it is they who made many of the discoveries that contribute to our understanding of the complexity of vision. Our work is a tribute to this long, painstaking, and exciting exploration. Shadows give us a window on vision, and this window has been found and opened by the work of artists and practitioners over the centuries. The main discovery is that a reduced set of features, a naïve physics of light and shadow, allows rapid processing of the visual scene by our visual system. But speed is not the only advantage of this simplified set of features, it also supports much of the richness of our visual culture. Specifically, the shortcuts used by our visual system allow artists to take the same shortcuts. Artists can take many liberties in depicting scenes that appear realistic without having to slavishly follow all the laws of physics. If, instead, our visual systems had to check all the physical properties of shadows, as is often the case in computer vision and computer graphics, they would never arrive at appropriate shadow labeling within a practical amount of time. And artists, in turn, would be constrained to strict photorealism in their representational pieces.

As we are discussing art, it is worth mentioning that we say little in this book about the fascinating cultural and symbolic aspects of shadows, the role shadows have played as actors in popular culture and in literature. These topics are covered in other recent works. But surely these aspects can exist only because the visual brain passes on shadow representations to other systems; one way or another, one needs to consider shadows for vision.

This is a science book. It tries to capture phenomena and provide explanations at the frontier of our knowledge. Inevitably, the science part will soon be outdated; this is the essential nature of science. Be that as it may, one of the hopes we have for this book is that the reader not only enjoys the rich images of shadows it presents, and glimpses the underlying mechanics, but at some point drops the book and starts looking around, chasing interesting shadows, in the world, in art, in movies. We wish to enrich readers' experience of the visual world; in a sense, ours is also a plea for observation. We can testify to the pleasure we have experienced in contemplating artworks in museums all over the world. No matter how well produced, photographic reproductions of artworks always leave something wanting, and our book is no exception—one more reason to go to the museum: a pleasure matched by that experienced in looking at light-and-shadow phenomena in nature, making for oneself all the small discoveries that added to our general picture. Observation takes many forms. It may be passive and slow, an exercise in patience; but at times it is also an exercise in fast decision making: that shadow on the sunlit wall will not last long. Quick, go outside and look!

Acknowledgments

So many people have helped us, it is simply impossible to name them all.

Special thanks from Roberto go to Francesca Bizzarri, Colleen Boggs, Fabrizio Corneli, Hannah Dee, David Freedberg, Anna Gialluca, the late Vittorio Girotto, Judith Haziot, Seiko Hoshi, Larry Kagan, John M. Kennedy, John Kulvicki, Paolo Legrenzi, Pascal Mamassian, Mario Martinelli, Robert Mougenel, Goffredo Puccetti, Valentina Rachiele, William Sharpe, Roy Sorensen, Paul Taylor, Koichi Toyama, Barbara Tversky, Achille Varzi, and Werner Weick. Thanks also to Paulo Santos and the participants at the 2015 Dagstuhl Shadow Seminar; to students at IUAV, Venice; at ENSCI, Paris; at CogMaster, EHESS ENS P5; at the Liceo Artistico Statale "Nanni Valentini," Monza; and at the seminar "Cognitive Artifacts," EHESS, Paris.

Roberto would also like to thank his family: Beatrice, Lise, Anni, and Nina Casati. Thanks are also due to the institutions and grants that made this work possible: Institut Jean-Nicod, Paris; the Leslie Center for the Humanities at Dartmouth College; the Italian Academy of Columbia University; the *soutien à la mobilité internationale* of CNRS; grants ANR-10-LABX-0087 IEC and ANR-10-IDEX-0001-02 PSL; Grant BQR-Renaissance of École normale supérieure, Paris; Grant IRIS-PSL, and Grant PSL-Dial.

Thanks from Patrick go to Arielle Veenemans, Alan Gilchrist, Hany Farid, Caeli Cavanagh, Ryan Cavanagh, and Josée Rivest, for contributions to this project; and to Yvon Leclerc, John M. Kennedy, Cassandra Moore, Satoshi Shioiri, Pawan Sinha, Yuri Ostrovsky, and Ron Rensink, for sharing his initial interest in exploring shadows, starting long ago in 1989. And thanks to the other pioneers of shadow research, Ted Adelson, Al Yonas, Pascal Mamassian, Dan Kersten, and Ernst Gombrich, who kept the darkness alive; and of course to Roberto, who shares Patrick's passion and was willing to work on one more book about shadows after his influential *The Shadow Club* (2002). This work would not have been possible without the support of the European Research Council under the European Union's Seventh Framework Programme (FP7/2007–2013)/ERC grant agreement no. AG324070 and the Department of Psychological and Brain Sciences, Dartmouth College.

Thanks from both of us to copy editor William Henry, and to the team at the MIT Press: Philip Laughlin, Anne-Marie Bono, Judy Feldmann, Emily Gutheinz, Mary Reilly, and Janet Rossi, for their work on this complex project.

Names of painters have been standardized using the Union List of Artist Names Online of the Getty Research Institute (http://www.getty.edu/research/tools/vocabularies/ulan).

Image credits are given in full, except for those to the book's authors, who are initialed "RC" (Roberto Casati) and "PC" (Patrick Cavanagh), respectively. The figure captions contain a time stamp and location where possible to give an approximate idea of the general light conditions of the situation.

This book contains many images, but actually only a fraction of the many more we would have liked to publish. These come from four main *corpora*. First, experimental stimuli are designed to test the visual system in controlled conditions; when reprinted here, they do not correspond to the intended setting but give a fair idea of the experimental conditions. Second, we used photos from our respective collections, documenting decades of observations in the open. Third, we drew from a database of 1,500 drawings of artworks that Roberto made in museums, pointing to aspects of shadow depictions that required *de visu* observations; work on the database was funded by the BQR Renaissance grant of École normale supérieure, Paris, and involved the collaboration of Valentina Rachiele and Judith Haziot. None of the drawings is published here, but—fourth—we used a number of images from museum collections; Patrick's grants made it possible to secure rights and to enlist Arielle Veenemans for the documentary and administrative steps. We are extremely grateful to all the institutions around the world that authorized us to reprint copyrighted images in our volume and to Wikimedia for the free access to a wide range of images.

A remark about image copyrights is perhaps in order. As an art historian colleague once told us, half of his life was devoted to securing rights to publish images from various collections around the world. The time-consuming part of the process is absolutely prohibitive and, we daresay, not well spent. It can be surreal, too. A researcher may use funds from a grant by her government to pay for the rights to publish an image from the collection of a state museum in her country: the grant money thus leaves, and then returns to, the state's treasury—a lose-lose for the taxpayer, given the time spent by the researcher in working on both the grant and the rights. The situation has improved a bit since. Some institutions, which we wholeheartedly thank, waived fees but still required us to go through some administrative steps. Everything will change when institutions align to the gold standard of the Metropolitan Museum of Art in New York and the Rijksmuseum in Amsterdam: use of a Creative Commons Zero license, that is, free publication in academic books for all the digitized images in their collection, with no other red tape than mention of the appropriate metadata. (There are known ways to get there: private donors can make themselves popular by paying wholesale for reprint in academic books—the sums involved being far from prohibitive.) Meanwhile, as our work is not about a particular artwork or artist but about pictorial solutions in general, every time we needed an example, we tended to turn to these noble institutions, sacrificing other pictures in other museums. We think that this situation is suboptimal, too. We hope that in the end, museums all over the world will understand that publication in academic books is a way to give value to their holdings, not a venue to be taxed, as would use for advertising.

In yet other cases, we performed what we may call the "Arnheim move": hand copying artworks, indicating where you can find the relevant shadow. (Here we take advantage of the large availability of images on the web and trust that the reader will search for them if needed.) Finally, in quite a few cases, we staged the depicted scene. Some photos have been digitally manipulated. When this has occurred, we signal it in the text or in the caption. Digital manipulation is motivated by the need to create contrasting stimuli. In general, open-air or staged pictures document actually occurring environmental light and shadow phenomena.

Most photographs by Roberto were taken with the camera embedded in his outdated cell phone, the Sony Ericsson W810i, bought in 2006. "It's an old model that I used until its death in 2015," Roberto says, "because it's not a smartphone, has a very low battery consumption, and takes nice pictures that are not very large (300k on average). I could have shot higher-definition pictures using a professional camera, but most of my pictures are taken in the moment, when the only resource was the phone in my pocket. One must be ready; shadows are so elusive. They have been a great school for exercising my visual attention, and low(er) tech is sometimes a good asset."

A note on color: Thanks to the MIT Press for accepting from the outset that this would be a full-color project. Shadows are dark, but they are not captive to a black-and-white world.

1 Information in Shadows

1.1 Background

Shadows are holes in light, created by an opaque or semiopaque object blocking the light falling on a surface. In this chapter, we describe the vast range of information that can be recovered from shadows, details well beyond what the human observer sees in any given image.

Everything that we see or that can be optically recovered is, of course, carried by structured patterns of light: more light here, less light there, lights of different wavelength, changes over time, and interruption in light flow. Shadows punctuate this pattern with their presence, shapes, and movements. They are a witness to the encounter of light with one or more objects, and when they fall on a surface, they reveal what they encountered. We will see that shadows can provide information about existence: of an object, of a surface, and of the sources of light. They further inform us about the shapes and positions of the objects casting the shadows, the location of the light, and the relief of the surface: if you know (or can visually access) two of these, you are often able to compute the third. By looking at shadows, we get a second point of view on the scene where the light is the observer and the shadows are the regions that observer cannot see. In addition, shadows can provide information about the quality of the light and the atmosphere through which it travels.

In this chapter, we primarily review not what the visual system *actually* exploits in cast shadows but what is there that it *could* exploit (although it may not be well equipped to do so). We outline the various types of information that *can* be extracted from shadows by systems that are significantly like ours, including artificial systems. These points are inherent in the details that emerge in computer graphics when rendering shadows in

Figure 1.1
Mount Rainier casting its shadow upward onto the clouds as the sun rises. Image credit: Kim Merriman, *Mount Rainier from Eld Inlet*, December 31, 2012.

scenes. Many are just the obvious: if there is a shadow, there is light. This chapter provides the basics of the ecological optics of shadows.

1.1.1 Facts about Light Constrain the Information in Shadows

In the eleventh century, scientist Ibn al-Haytham (Alhazen) demonstrated that light travels in straight lines and laid down the basic geometry that constrains the shape and location of shadows. When there is a single illuminant, each object creates a zone where that illumination is blocked. The unilluminated region on the object itself is called the *attached shadow*; the regions of blocked illumination projected onto other surfaces along the direction of the illumination are *cast shadows* (fig. 1.3).

It is unusual to have only a single source of light in a scene, as even this single source will reflect off all the surfaces on which it falls, creating secondary reflections that combine into a more diffuse ambient light. In outdoor scenes, the scattered blue light of the sky or diffuse light from cloud cover add to this ambient light.

When the light source is small and illuminates a nearby object (fig. 1.4), the diverging rays strike the object and cast a shadow cone, or *shadow body*.

Figure 1.2
The sun casts complex sharp shadows of the bike on the ground and wall. Image credit: RC, Pisa, October 2015.

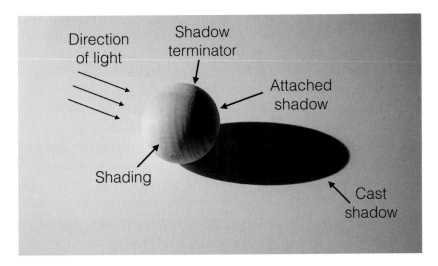

Figure 1.3
Some terminology: The amount of light reflected from the illuminated portion of the sphere's surface on the left varies with its angle relative to the illuminant, and these variations in brightness are called *shading*. The edge between the illuminated portion of the sphere and the portion not directly illuminated is the *shadow terminator*. The portion of the sphere not receiving direct illumination is in shadow, and this is the *attached shadow*. Finally, the sphere projects a *cast shadow* onto all surfaces that it shields from the direct illuminant. Image credit: PC.

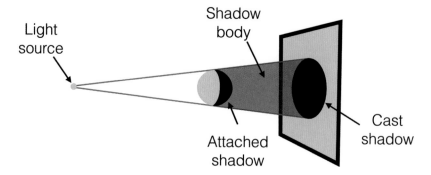

Figure 1.4
A point light source casts a sharp shadow on the far surface and leaves the back side of the sphere unilluminated (an attached shadow). Between the sphere and its cast shadow is a volume of unilluminated space, the shadow body. The shadow body is not normally visibly darker, as depicted here for convenience. It is darker only if there is fog, mist, or smoke in the air to make the illuminated volume outside the shadow body lighter than the unilluminated volume within it. For examples, see figures 2.1 and 2.2. Image credit: PC.

An object will cast shadows of different sizes depending on its distance from the surface onto which the shadow is cast. In this simplest case, the size of a cast shadow can tell us about its distance from the object that casts it and the distance to the light source.

However, light sources are often extended, whether they are the sun, diffuse light from cloud cover, or artificial light. These broad light sources cast shadows that become increasingly faint as they fall farther from the object blocking the light (fig. 1.5). Extended sources can be modeled as a large number of point-like sources, each casting shadows that fall on areas that also receive light from other points in the source. In this case, a shadow is typically only noticeable on surfaces right next to the object that casts it (fig. 1.6, left).

The sun itself is not a point source. It covers about 0.5 degrees across in the sky (about 360 suns could be placed across the sky from horizon to horizon, each just touching the next). On the Earth, light from each point of the sun has effectively parallel rays, but given the size of the sun, the edge of an object's shadow will display a gradient from light to dark as more and more of the sun is blocked by the intervening object. The width of the resulting penumbra indicates the distance between the object casting the

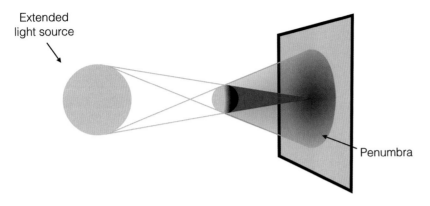

Extended
light source

Penumbra

Figure 1.5
When the light source is extended, it casts a diffuse shadow where the object partially blocks the light. The attached shadow on the back side of the sphere remains completely blocked from the light source, but the far surface shows a wide penumbra where the light changes from being unimpeded at the edge to partially blocked in the middle. Image credit: PC.

shadow and the surface receiving it (see sec. 1.2.5). For light sources other than the sun but at a fixed distance from the object, the width of the penumbra changes with the size of the light source (fig.1.6).

Parallel or diverging light rays are the norm, but light rays can converge in special circumstances, such as after passing through lenses or after reflecting from concave surfaces. Rippled surfaces in water can generate optical shadows, that is, patches of dimmed illumination without the presence of any object casting a shadow (fig. 1.7, left). We can see a similar effect in reflections from an irregular, discontinuous surface, where dark areas look like cast shadows but are caused by deflecting or absorbing the light into a hole, and not by an intervening object (fig. 1.7, right). Similarly, when light is refracted or expanded from a lens, it leaves a diminished illumination within its frame (fig. 1.8).

1.1.2 Shadows and Projections
Mathematically, cast shadows are projections (fig. 1.9). Consider a square casting its shadow onto a flat screen underneath. If the source of light is at infinity and rays are parallel, we have a metric projection whenever the square lies in a plane parallel to the plane of the screen (topological, metric, and angular two-dimensional properties of the square are preserved by its

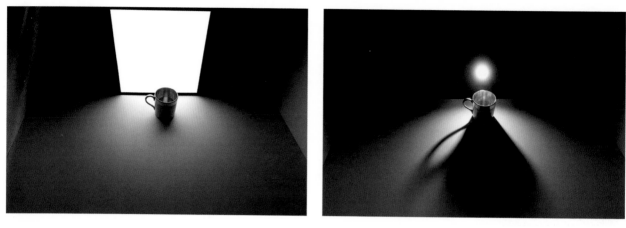

Figure 1.6
A more extended light source (*left*) creates a more diffuse penumbra at the edge of the cast shadow. Image credit: PC.

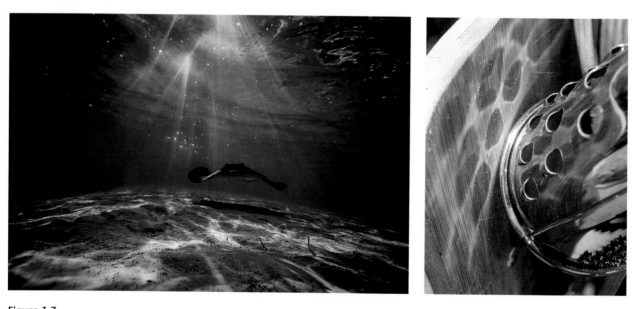

Figure 1.7
Shadowlike light effects that are not shadows. When the surface of the water is not flat, light is bent in many directions, creating local summations and absences that are called *caustics* (*left*). Similarly, if a reflective surface is irregular or discontinuous (*right*), it carves holes in light that are not shadows. Image credits: Left: Reed Arce, *Sun Rays* (reprinted with permission). Right: RC, Paris, 2015.

Figure 1.8
A pair of prescription glasses expands the light passing through them, creating a reduction within the frames, perceived as a shadow; and a summation, perceived as a light spot. Image credit: RC, Paris, November 2015.

shadow; see fig. 1.9a), as a limit case of an affine projection whenever the square does not lie in a plane parallel to the plane of the object casting the shadow (parallelism of sides preserved, but metrics not preserved; see fig. 1.9b and c).

If the (point-like) light source is moved close to the square—which in turn is parallel to the screen—we obtain a similar projection (the shadow is just a larger square than the object that casts it; angles preserved, metric properties not preserved; see fig. 1.9d).

If, while keeping the source at a finite distance, we tilt either the square object or the screen but keep one side of the square parallel to the screen, we have a more generic central projection (fig. 1.9e); and if we tilt either the square object or the screen and keep no side of the square parallel to the screen, we obtain an even more generic central projection (fig. 1.9f) where the metrics and parallelism of the sides are not preserved, but ratios of distances between opposing vertices are.

Figure 1.9 shows a simplified scheme to convey the basics of shadow geometry. As we said, some underlying assumptions are at work. The square is a two-dimensional opaque entity, but in real life, objects that cast shadows are three-dimensional and may be more or less opaque. Interactions

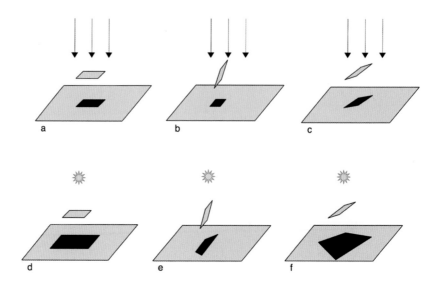

Figure 1.9

A square casts its shadow onto a flat screen. *Left column*: plane of the square parallel
to the plane of the screen. *Middle column*: square tilted with two sides of the square
parallel to plane of screen. *Right column*: square tilted with no sides of the square
parallel to plane of screen. *Top row*: source at infinity. *Bottom row*: source close. Projection of the shadow is metric (*a*): the shadow has the same shape and size as the
square; affine (*b–c*): the shadow is either a rectangle or a rhomboid, only parallelism
is preserved; similar (*d*): the square projects onto a shadow of different side lengths,
angles are preserved; projective (*e–f*): ratios are preserved. Image credit: RC.

between the material and the light are not taken into account, but in real
life they may complicate the pattern. The source of light in the bottom
row is considered to be point-like. In either case, penumbra effects are not
considered.

Because shadows are projections, the shape of the projection will depend
on the shape of the object, the position of the projection point or points
(the light source), and the relief of the surface on which the projection is
cast. The same object will cast dramatically different shadows when one of
the parameters is significantly altered (fig. 1.10).

1.2   Basic Information in Shadows

In this section, we present several basic aspects of the information contained in shadows. Shadows may reveal the existence, size, position, and

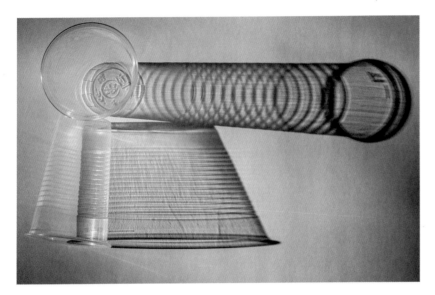

Figure 1.10
Shadows of two identical plastic cups. The different orientations of the cups relative to the projection surface result in very different shadows—to the point that you may be surprised to discover that they both are shadows of the same object shape. The top cup is right side up. Image credit: RC, Paris, March 2017.

shape of objects; the presence and topology of a surface receiving the shadow; and the presence, direction, and color of a light source. This list of the information contained in shadows is not exhaustive, and we have organized the text to best guide the discussion of shadows in subsequent sections rather than following a formal presentation of geometric optics of shadows. In chapter 2, we describe a set of secondary aspects of shadows: the shapes of shadow bodies, ambiguities of shadows and silhouettes, inter-actions of shadows and scene details, artifacts of shadows, and movement in shadows.

1.2.1   The Existence of Objects
The pattern of light from the scene specifies some properties of the objects that we perceive: color, shape, location, and movement—as well as other, more complex properties, such as *affordances* (what we can do with or to the object, such as sit on it, grasp it, throw it), or friendliness, fearfulness,

or even intentional mental states. In all these cases, some complex structure in the visual array correlates with the properties, and it is the business of the visual system to take advantage of the correlation to compute the property. Many of these properties are, however, relatively difficult to recover. A much more basic point appears not to have attracted much theoretical attention, namely, bare *existence*. In the case of shadows, the presence of a shadow informs us of the existence of an object casting it and of a relatively circumscribed source of light. Moreover, by giving access to items that are not visible, shadows can provide information about otherwise hidden objects. And by providing a viewpoint that is not accessible from one's current viewpoint, they can reveal the structure of parts that are not in view. Here we consider the implications for image regions that actually are shadows; in chapter 6 we consider how to determine whether an image region is or is not a shadow.

Let us state the first and simplest of the ecological laws of shadows (others will follow):

If there is a shadow, there is an object that casts it.

The connection generates expectations. You spot what you believe to be a shadow (not a stain or pigment, etc.). You may be able to see the object that casts the shadow, or you may not. But you will often feel a need to search for it. Movie directors, painters, and photographers have used shadows to generate expectations of presence or to widen the content of a picture to something that cannot be contained in its frame (figs. 1.11, 1.12).

Forensic literature uses shadows in photographs as cues to the existence of objects that are not visible in the picture. Pictures may be massaged to make some items appear or disappear, but then one should remember to keep the shadow properties appropriate for the illumination and layout of the scene.

Although it is true that every illuminated object casts a shadow (though it may not always be visible), the reverse, that every shadowlike dark region must have an object that casts it, is not strictly true. With multiple light sources, it is relatively easy to construct an apparently illuminated field surrounding an unilluminated region. This may look like a shadow, perhaps intentionally so, but there is no object creating the absence of light in the region. This is the case for caustics and reflective surfaces that themselves

Figure 1.11
We know that a picture frames only a portion of the scene and we know that there is something outside the frame. Shadows make us *see* that there is something else around. Hermann Eichens (1813–1886), engraving with chisel and aquatint, after Jean-Léon Gérôme (1824–1904), *Consummatum est!*, 1867. Color print, 1871. Bordeaux, Musée Goupil, inv. 97.I.2.125. Image credit: © Mairie de Bordeaux, reproduction photo B. Fontanel.

Figure 1.12
An object overhead. Planes landing in Sint Maarten must cross over the nearby Maho Beach at extremely low altitude. Image credit: Thomas Prior, Pinterest.

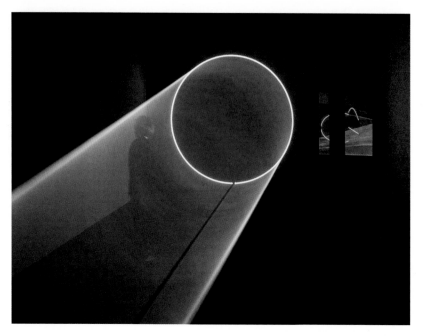

Figure 1.13
A dark region surrounded by light is not necessarily a shadow. Here it is produced
by manipulation of light sources. Anthony McCall, *Line Describing a Cone 2.0*, 2010.
Installation view, LAC, Lugano, 2015. Image credit: Photograph by Stefania Beretta.

have holes in them (recall fig. 1.7) and for holes in a distribution of light
constructed by projection (fig. 1.13).

1.2.2   The Existence of Surfaces and Their Topology

A shadow reveals more than the presence of an object. It also reveals the
presence of a surface—the surface onto which the shadow is cast (fig. 1.14).
The shadow may or may not reveal much about the properties of the surface, but it definitely tells us that there *is* a surface.

If there is a shadow, there is a surface it must be falling on.

The *shape* of a shadow is the intersection of the shadow body, the region
of space where the object has blocked the illuminant, and a surface on
which the shadow becomes visible. A shadow's shape is modified by the
layout of the surface on which it falls. If it falls on a flat surface, the shadow's shape is some more or less stretched version of the object's outline (as

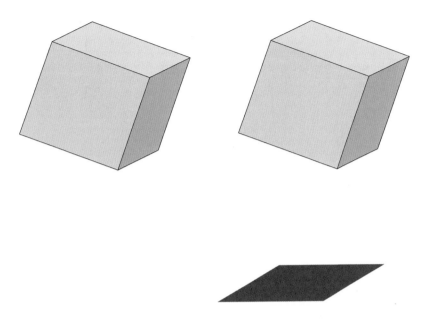

Figure 1.14
The scene on the left contains a cube, but we cannot say much more than that. The image on the right tells us something more: The scene contains a cube *and* a surface. The surface appears to be flat in the area where the shadow is cast. We do not know if the surface is only slightly larger than the shadow, or whether it reaches out to the horizon. But we see that there *is* a surface. Shadows can make surfaces visible and hence reveal their existence. Image credit: RC.

depicted in fig. 1.9). If the receiving surface has a complex topology, the resulting shadow is complex, and the object's shape and the surface topology cannot be derived simultaneously based on the shadow alone. We can make some explicit deductions, though. For example, in figure 1.15, the railings on the left are visible and continuous, whereas their shadows are discontinuous. This indicates that the surface has discontinuities in depth.

Although in general we cannot infer the geometry of a complex surface from the shadow's shape alone, the inference may become possible if the surface or the object or the light source is moving. If the object casting the shadow and the source remain constant, then changes in the shadow on a moving surface can reveal its local geometry. A shadow falling on a moving vehicle will appear to bend, shrink, or expand, signaling edges, hollows, or bumps, respectively. However, informal observations suggest that even

Figure 1.15
When falling on complex, articulated surfaces, shadows themselves are quite complex. The pattern here indicates a regular discontinuity of the surface that breaks the shadow of the continuous casting object. Image credit: PC, Hanover, New Hampshire, April 18, 2017.

though the recovery of the surface should be possible, perception may not be especially attuned to this type of information and may simply stop at the interpretation that the shadow behaves in a strange way.

1.2.3   The Presence and Location of Light Sources

It takes three elements to make a shadow (light, object, and surface), but a light source is definitely mandatory. No light, no shadow. Inverting that logic, if there is an actual shadow, there must be a light source around. Shadows thus necessarily reveal the presence of light sources in the environment.

If there is a shadow, there is a light source.

Light sources, on the other hand, do not automatically create discernible shadows. That is the case with extremely diffuse light (e.g., on a cloudy day) where shadows are only evident on surfaces very close to an object. Or consider a different environment, inhabited by objects that are luminescent, as is the case (weakly) for many deep-sea creatures, or in some discotheques. The objects would be visible, but shadows, if any, less so. Even in our environment, lights and objects may be distributed so as to create shadow noise, that is, a large number of shadows that overlap, such that recovering the correspondence between objects and their shadows becomes difficult.

Nevertheless, when shadows are well defined, the pairing of the object and shadow features can be put to work to retrieve the position of the light source (fig. 1.16).

A light source is at the end of the solid angle that spans both the cast shadow and the object casting the shadow.

The light source is at the intersection of lines in the scene connecting points on the shadow to points on the object that cast them. Here we label these lines *shadow rays*. In an image, the shadow rays converge on the source if it is in front of the observer, as it is in figure 1.16 and figure 1.17 on the left. In contrast, when the source is behind the observer, perspective effects will make the shadow rays converge on what we can call the antisource. For example, on the right in figure 1.17 and on the left in figure 1.18, the light–shadow boundaries converge not on the sun in the image but on the point opposite to the sun. The sun is behind our back, its rays are nearly parallel to the ground, and if we followed the light–shadow boundaries in the scene

Figure 1.16

Recovering the position of the light source. *Left*: When lines are traced in the image from individual shadow features through the object features casting them, these "shadow rays" will converge on the source if it is in front of the observer, as it is here. *Right*: In other cases, the source, even though in front of the observer, may be out of the image frame. The shadow rays are in red, and the source, the location of the sun, is shown as a red disk above the image. The central axes of shadows of vertical objects on horizontal planes (shown as orange lines) also converge in the image, but to the vertical projection of the source onto the horizontal plane—in the case of the sun, at the horizon (orange disk). Image credit: PC.

Figure 1.17

Left: Rays from the sun at sunrise or sunset are (practically) parallel to the Earth's surface, but because of perspective, they converge to the source in the image when it is in front of us. *Right*: Rays converge to the antisolar point when the sun is behind us. Image credits: Left: RC, Barbizon, France, August 2013. Right: RC, North Atlantic, January 3, 2017.

back over our head and beyond, we would find the sun. But in the image, the shadow rays must converge to the antisun, or the antisolar point.

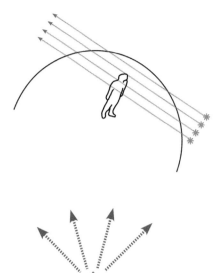

Shadow rays in the image converge to the source when it is in front of the observer, converge to the antisource when it is behind the observer, and are parallel if the source is in the image plane (like a flash attached to a camera).

Shadows of mostly vertical objects (people, posts, tree trunks) provide an additional set of cues regarding the light source location when they fall on flat ground, as in figure 1.16 on the right (orange traces).

The axes of shadows cast by vertical objects on a horizontal plane converge in the image to the vertical projection of the source on that plane, if the source is in front of the observer, or to the projection of the antisource to the horizon along the direction of illumination, if the source is behind the observer.

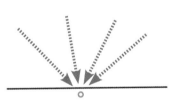

These properties of shadow rays and shadow axes provide a pivotal check for consistency of illumination in an image. They can be used to detect poorly altered images (see chap. 8), but these alterations often go unnoticed by human vision (see chap. 6).

In figure 1.18, left, the position of the antisun further *specifies the observer's position* relative to the shadow of the mountain. We can thus infer that the picture was taken from close to the top of the mountain. Shadow rays tell us the line along which the sun lies even when it is not visible. For example, in figure 1.18, right, we cannot know the sun's position along the line joining the Matterhorn's peak to its shadow, as the sun lies behind the mountain. We know only that it is still high enough to illuminate the other side of the mountain.

When the sun is close to setting or has just risen, rays are parallel to the ground (*top*). Because of perspective, they are seen to converge toward the sun (*middle*) or toward the point opposite to the sun, the antisolar point (*bottom*).

1.2.4 The Position of the Object

Shadows can provide the relative position and distance of objects and surfaces. The basic constraint is topological and involves contact or, in the more informative case, *noncontact* or *detached shadows*.

If, in the image, the shadow does not touch the object that casts it, then in the scene the object does not touch the surface on which the shadow is cast.

Figure 1.19 nicely makes the point that we can use shadows to estimate the distance of an object from the surface on which its shadow is cast. In this case, the perceived distance is more or less accurate relative to the bottom of the sea (modulo refraction and estimating boat size and elevation of

Figure 1.18

Left: View from Mount Fuji. Here the sun is at our back, and the shadow rays appear to converge to the antisolar point. Note that the convergence of the rays to the top of the shadow of Mount Fuji reveals that the picture was taken from a location close to the top of the mountain. *Right*: The Matterhorn dramatically doubles as its shadow is cast on clouds. The sun's position must lie along the shadow ray joining the top of the mountain to the top of its shadow. Image credits: Left: Image credit: Mount Fuji, September 14, 2012, 18H. Meru Kukichi, https://youpouch.com/2012/09/14/81731/. Right: RC, Zermatt, February 2013.

Figure 1.19

The boat appears to float above the surface of the sea. Image credit: Danos Kounenis, Skiathos, Greece.

Figure 1.20

Left: An anchoring shadow. The cube is sitting on the surface in the scene, and the cast shadow touches the cube in the image. *Middle*: The cube is not touching the surface in the scene, but its cast shadow still touches the cube in the image. *Right*: A detached shadow. The cube is above the surface, and the cast shadow does not touch the cube in the image. Image credit: PC.

the sun). However, since no pictorial cues link the boat to the sea surface, it looks as if it is floating in air, above the sea surface.

What can we infer from cases in which the shadow does touch the object in the image? These contact cast shadows often serve to anchor the object to the surface (fig. 1.20, left), but this contact may be accidental. An object that does not touch the surface on which it casts a shadow may be aligned so that it overlaps it in the image from some viewpoints (fig. 1.20, middle).

Despite this loophole, contact shadows typically convey physical contact in the scene and are widely used in art to anchor objects to the surface below them; conversely, cast shadows that are detached are widely used to make an object appear to float (fig. 1.19 and fig. 1.20, right). Nevertheless, in the visual image, contact between object and shadow may be accidental, so we should phrase this constraint as a consequence of contact, not as an opportunity to infer contact.

Further constraints help resolve these indeterminacies. On the left in figure 1.20, the shadow on the ground touches the borders that cast it, forming a continuous outer boundary joining the shadow and the object. In the middle, the shadow contacts the object in the image, but its contours are not continuous with those of the object. This indicates that the shadow is not in contact with the cube, even if the image of the shadow is in contact with the image of the cube. This constraint of object–shadow continuity provides one way to disambiguate distance and contact (see chap. 4).

If an object touches a surface on which it casts a shadow, the shadow contours must mate with the object contours.

1.2.5 Penumbras and the Distance between Object and Surface

Lights that are point sources are rare (e.g., a compact camera flash), but for these sources, the edge of a shadow is sharp no matter how far the surface onto which the shadow falls. More commonly, light sources are extended, like the sun, a fluorescent light, or a cloud layer that may cover the whole sky.

In the case of illumination by an extended light source, the width of the shadow's blurred edge—its penumbra—is proportional to the spatial extent of the light source and to the distance between the object casting the shadow and the surface receiving it (fig. 1.21).

For a fixed light source, the wider the penumbra, the greater the distance between the object casting the shadow and the surface on which the penumbra appears.

Interestingly, sundials face unsolvable engineering problems due to this link between distance and penumbra width. To make observation of the

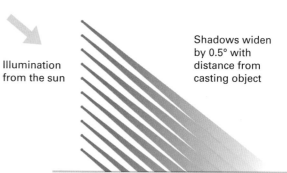

Illumination from the sun

Shadows widen by 0.5° with distance from casting object

Figure 1.21
Because the light from the sun is not a point source, the penumbra widens with the distance between the object and the surface. *Left*: The shadows of the fence grid are sharp for portions near the ground but grow progressively blurred and overlapped for the higher portions farther from the ground. *Right*: Penumbra widens with distance from the wires in the fence. The angle is exaggerated here. Image credits: Left: PC, Hanover, New Hampshire, November 2, 2017. Right: PC.

Figure 1.22
The 27-meter-high gnomon of the Samrat Yantra sundial of the Jantar Mantar in Jaipur, India, was made very tall to increase the travel of its shadow from dawn to dusk,
theoretically giving a more precise indication of the time. Unfortunately, because of
the large distance from the gnomon to the projection surface, its shadow is blurred,
negating the increase in precision. Ivars Peterson, *Jantar Mantar Sundial, Jaipur, India.*
Image credit: mathtourist.blogspot.com.

time from the gnomon's shadow as precise as possible, one may want to
increase the size of the whole instrument. In the case of the Samrat Yantra
(giant sundial) of the eighteenth-century Jantar Mantar observatory in Jaipur (fig. 1.22), this size race produced a twenty-seven-meter-tall gnomon,
and markings on the projection surface that in theory could indicate time
intervals as short as a half second. However, the great size of the gnomon
increases the distance to the surface on which its shadow falls, inevitably blurring it. In this case, the penumbra can be three centimeters wide,
largely offsetting the expected gains (we'll see a solution to this in chap. 6).

However, this widening effect can be visually nullified under some circumstances, most notably when the observer and the light source are on
the same side of the object casting the shadow. In this case (fig. 1.23), the
penumbra will still get wider with distance from the casting object, but its
visual angle to the observer will not, as it is seen from farther away, so that

Figure 1.23
Left: A schematic representation of the shadow of a pole seen when the observer faces the sun, and then the shadow seen when the sun is at the observer's back, where the widening of the penumbra is canceled by the narrowing of the shadow with distance. *Middle, right*: two real-world instances of these two cases. Image credit: RC, Ury, France, September 2014.

it appears to be unchanging in width. The interaction of increasing penumbra width with distance from the object and decreasing visual angle with distance from the viewer is informative but potentially misleading. As an example of information available from the penumbra, if we know the spatial extent of the light source (in degrees of visual angle), we can infer the distance from the shadow to an unseen shadow-casting object by looking at the width of its shadow's penumbra (fig. 1.23).

The effect of the orientation of a penumbra relative to the surface it falls on creates interesting visual effects. The penumbras of the vertical poles in the fences in figure 1.24 are compressed because of the viewing angle, but those of the horizontal rails are not. The shadows of the poles appear to pierce through the shadows of the horizontal rails.

1.2.6 The Size of an Object

The recovery of an object's size from its shadow is possible under special circumstances. In the natural environment, the normal illuminant is sunlight (or moonlight). This results in parallel rays, on average, with some added penumbral effects. In the case where the projection plane is normal to the

Figure 1.24

Shadows of roadside fence posts and rails. In this view, foreshortening narrows the penumbra of the posts but not of the rails, creating the impression that the shadows of the posts pierce those of the rails. Image credit: RC, Barbizon, France, August 2017 and May 2015.

direction of light, this creates shadows whose size, ignoring penumbras, is about the same as the size of the object casting it. The shadow of a bird is in many cases not very different in size from the bird. The limiting case is that of a sphere that offers a round cross section from every direction of the incident light; it is projected onto a normal plane as a circle (fig. 1.25, left).

If the surface on which the shadow falls is flat but is not normal to the light direction, the sphere projects to an ellipse (fig. 1.25, middle). The longer axis of the ellipse is stretched on the receiving surface to cover more than the sphere's diameter, but the shorter axis is not (fig. 1.25, right). The sphere casts a cylindrical shadow body, and every cross section of a cylinder is an ellipse. The longer axis of the ellipse can be of any length, but the shorter axis must be the diameter of the cylinder—which is also the diameter of the sphere. By measuring the shortest axis of the shadow (in the scene, not in the image), we have a good idea of the diameter of the sphere that casts it. The general heuristic roughly holds for irregular bodies as well.

When the illumination rays are parallel, the length of the shortest axis of an object's shadow in the scene is identical to the length of one of the object's cross sections.

A sphere and its elliptical shadow. Image credit: RC, Lefkada, Greece, August 2017.

Figure 1.25

Left: A sphere projects a round shadow on a plane normal to the direction of light (which is seen as an ellipse in the image). If the light rays are parallel, then the circular shadow is identical in size to the cross section of the sphere defined by a great circle. *Middle*: The projection rays are still parallel, but they are not normal to the projection surface. The sphere projects to an ellipse. *Right*: In an affine projection, the length of the shortest axis of the shadow in the scene is equal to the diameter of the sphere. Image credit: PC.

Even in the case of extremely elongated shadows at sunrise or sunset, a cross section through the shadow that is normal to the direction of light will effectively represent the size of the object casting the shadow in at least one of its dimensions.

There are few natural cases in which light rays are divergent, not parallel, so that they create significant differences between object size and shadow size. Point sources in the environment come from reflections from curved surfaces, which are rare, or from fires. The only effective natural point sources are created by pinholes that filter out parallel rays in sunlight and leave only those that converge from across the surface of the sun to the pinhole and propagate past it (fig. 1.26).

In all the preceding cases, we have dealt with relative sizes of an object and its shadow. However, the apparent size of the object casting the shadow relative to other objects in the scene can be influenced by the position of the object's cast shadow. In figure 1.27, the two bottom spheres subtend the same visual angle and could be of any size: they could be small spheres close to the observer or large spheres farther away. Adding a shadow that locates them relative to the horizon constrains their relative sizes.

1.2.7   Shape of the Object from Cast Shadows

Shadows provide silhouette information about an object's shape. The information is especially useful because it is a different view of the object's silhouette, the one seen from the viewpoint of the light. Consider the left

Figure 1.26

Top: Ecological conditions for diverging rays. The pinhole in the foliage creates a conic light beam. Objects closer to the receiving surface will cast smaller shadows than objects closer to the pinhole. *Bottom*: Reflection from a curved mirror generates divergent solar rays, so that the hand's shadow grows larger as the hand approaches the curved mirror. Image credit: RC, Lefkada, Greece, August 2017.

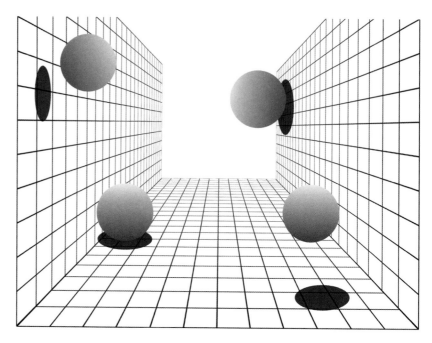

Figure 1.27
Shadows make these four spheres of the same dimensions appear to have different sizes. The shadow for the top right sphere places it significantly farther away, and it appears to be larger than the others. The two bottom spheres are at the same height in the image and placed at identical points relative to the vertical walls. Nevertheless, their two shadows place them at different distances in the scene, and the right-hand one (closer in the scene) looks smaller. Image credit: PC.

panel of figure 1.28, where we are facing a wire mesh fence. In this direct view, the mesh appears to be fairly flat, but the shadows indicate otherwise. They describe the object's shape as seen from the viewpoint of the sun. (The shadows could also be interpreted as indicating an undulating projection surface, but the absence of appropriate shading and highlight cues negates that interpretation.)

A cast shadow provides a second view of the object casting the shadow.

Taking advantage of the viewpoint of the light can solve image interpretation problems. Satellite views tend to flatten out shapes. Shadows provide extra information. In the right panel of figure 1.28, the aerial photograph of the Pont de Normandie is enhanced by the majestic shadow of the bridge

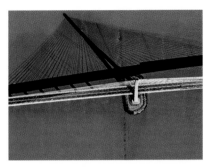

Figure 1.28
Left: Looking at the shadow, we have a better sense of the extent of the wire's bends. Image credit: Left: RC, Place de la Contrescarpe, Paris, March 2013. *Right*: Satellite photograph of the Pont de Normandie, France. North is on the left in the picture. Google Earth, retrieved on November 7, 2009.

on the water, revealing many details about the suspension cables and supporting columns not seen in the top view of the bridge.

If shadows reveal something about what can be seen from the viewpoint of light, they generally do so in a distorted way. The work of the shadow sculptor Larry Kagan exemplifies this point. Kagan works with wire to produce complex, nonrepresentational sculptures attached to walls. The parts of the sculpture are so disposed that their shadows, cast from a suitably placed lamp, merge to create something like a line-drawing representation of an object (fig. 1.29, left).

You can put your eye where the light is—or close enough to that position—and that is the only position where the shadow matches the visual profile of Kagan's artwork (a "copycat" shadow; see chap. 5). However, from that position the shadow disappears, hidden beyond the material part of the object (fig. 1.29, right). Think of the sun, which can never see the shadows it casts. With the exception of objects that have certain rotational symmetries (like cylinders), the visible profile of the object is different from the visible profile of the shadow seen from any other position.

1.2.8 Shape of the Object from Attached Shadows

Attached shadows are powerful indicators of the concavity and convexity of a surface. This is one of the main uses of shadow representations in graphic works—in art, in design, and in digital rendering. A general rule is available here:

Figure 1.29
Larry Kagan, *Frankfurt Chair*, 2002. *Left*: The jumbled wire casts a shadow in the shape of a chair. *Right*: An image taken from a position close to the light source reveals that the arrangement of wires does indeed take the shape of a chair from this viewpoint, and its shape hides its same-shaped shadow from view. (The large shadow is from the hand holding the camera close to the light source.) Image credit: RC, Troy, NY, September 2013.

Moving in the direction from the light source, a convexity produces a light-then-shadow pattern; a concavity produces a shadow-then-light pattern.

As an example, consider a small hemispheric bump in front of you, lit up from the right-hand side (fig. 1.30). Light encounters the bump on its right-hand side, which shows up brighter in the image, while a shadow forms on its left-hand side. If the position of the light is kept constant and the pattern of brightness and darkness inverted, there is now a depression in the surface—and no longer a bump.

At the informational level, the disambiguating factor here is the direction of light. Absent information about the direction of light, the light–dark pattern on a surface is ambiguous. We will see (chap. 4) that the visual system makes an assumption about the direction of light: light typically comes from above, and the visual system has adapted to this circumstance.

Figure 1.30
Left: A real-world bump. *Right*: Cross section sketch of a surface lit from the right-hand side. Relative to the direction of light, concavities (on the left) will present a shadow-first-then-light pattern whereas convexities (on the right) will present the opposite. Image credit: RC.

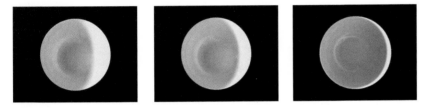

Figure 1.31
Three phases of a bowl (light from the left) that are inverted relative to phases of the moon (which would have light on its right). Image credit: RC.

Not taking the direction of light into account may have been responsible for some misconceptions. Heraclitus (ca. 540–480 BCE) claimed that the moon is a bowl of which we see the hollow part, as in figure 1.31. This interpretation of the shape of the moon is incompatible with the relative positions of moon and sun in the sky, but the image of the moon on its own does not specify which interpretation is the correct one. Similarly, Ptolemy (ca. 100 CE) noted that a distant sail could sometimes be seen as convex and sometimes concave even though, given the direction of sunlight, only one interpretation is possible.

In addition to indicating whether a surface is convex or concave, the attached shadow border reveals a single contour along the object's surface that follows the path where the light just grazes the surface. This can give us a powerful source of shape information when the attached shadows intermingle with the visible object boundaries. For example, we can reconstruct the three-dimensional faces of the two men and the woman in

Figure 1.32
Mooney faces. Dark areas are interpreted either as black pigment or shadow. Light areas can be interpreted as either light pigment or reflection. Image credit: PC.

figure 1.32 despite the extreme sparseness of the image information. To do so, we give quite different interpretations to otherwise identical dark areas: some are seen as dark pigment, like hair, and others as lighter pigment, like skin, in shadow.

1.2.9   Color and Quality of Light Sources

Most types of information we have covered so far concern *geometric* properties of the scene. However, shadows also reveal qualitative aspects of the world. In particular, they can indicate features of the light such as intensity and color. The intensity and shape of the light source determine the contrast and sharpness of the shadow border. The stronger and more point-like the light, the sharper the contrast at the border.

Shadows also act as litmus paper for the color of the ambient light. The shadow is protected from direct light and thus reveals reflected or atmospheric light. Blue skies confer a blue tone to the shadows (fig. 1.33, left).

The color within a shadow reveals the color of the light sources not casting the shadow.

Figure 1.33

Left: Under cloudless, sunny skies, shadows are filled with bluish skylight. *Right*: In the presence of multiple sources of light, each shadow reveals the color of the light that is not casting the shadow. Image credits: Left: Emanuela Zilio. Right: PC.

In general, when multiple light sources of different colors are present, the colors of the sources mix additively at any nonshadowed point, and retrieving each source's color becomes an intractable problem. However, if the light sources are at different locations, each casts a separate shadow that is filled by the other sources (fig. 1.33, right).

1.3 Summary

We have seen that patterns of light and shadow contain a wealth of useful information that biological or artificial visual systems can use to reconstruct basic features of the perceived scene: sheer existence; relative distance of objects; shape of objects and relief of surfaces; size of objects; presence, nature, and position of light sources; and viewpoint and perspective. The extraction of these properties from a given shadow pattern may depend on several constraints that help solve a reverse optics problem, but in most cases, it is possible to specify the constraints and limit the class of solutions. Here is a list of simplified constraints that we have discussed so far for

The shadow cast by the red light is green and vice versa.

regions that actually are shadows (we consider the problem of determining what is or is not a shadow in chaps. 5 and 6).

If there is a shadow, there is an object that casts it.

If there is a shadow, there is a surface it must be falling on.

If there is a shadow, there is a light source.

A light source is at the end of the solid angle that spans both the cast shadow and the object casting the shadow.

Shadow rays in the image converge to the source when it is in front of the observer, converge to the antisource when it is behind the observer, and are parallel if the source is in the image plane (like a flash attached to a camera).

The axes of shadows cast by vertical objects on a horizontal plane converge in the image to the vertical projection of the source on that plane, if the source is in front of the observer, or to the projection of the antisource to the horizon along the direction of illumination, if the source is behind the observer.

If, in the image, the shadow does not touch the object that casts it, then in the scene the object does not touch the surface on which the shadow is cast.

If an object touches a surface on which it casts a shadow the shadow contours must mate with the object contours.

The wider the penumbra, the greater the distance between the object casting the shadow (or a part thereof) and the surface on which the penumbra appears.

When the illumination rays are parallel, the length of the shortest axis of an object's shadow in the scene is identical to the length of one of the object's cross sections.

A cast shadow provides a second view of the object casting the shadow.

Moving in the direction from the light source, a convexity produces a light-then-shadow pattern; a concavity produces shadow-then-light pattern.

The color within a shadow reveals the color of the light sources not casting the shadow.

These constraints concern the information that is available from shadows. However, this information is not necessarily used by biological visual systems. And even if it is used, its use is not necessarily ideal or even optimally correct. In particular, given some reliable piece of information contained in shadow patterns, one or more of the following situations may occur, which we explore in this and subsequent chapters:

• The information is not exploited (e.g., pinholes in canopies; see fig. 1.26 and chap. 2).

• The information is exploited but is given only weak weight and can be trumped by other, conflicting sources of information about the scene (chaps. 4, 5, 6).

• The information is exploited only partially given environmental constraints (e.g., light from above; see chap. 4).

• The information is exploited incorrectly (e.g., illusion of position from shadows; see chap. 4).

The exploration of these mismatches between available and utilized information is fascinating territory for cognitive science and neurophysiology, as it points to idiosyncratic codes and mechanisms of the brain for dealing with visual information. It also suggests that there is room for the development of artificial systems, both for remediating and for augmenting vision by tapping into information in shadows that is neglected by biological systems.

Shadows in fog. Image credit: Michael Melford.

2 Secondary Aspects of Information in Shadows

The previous chapter outlined the basic information we can retrieve from shadows: position and shapes of casting objects, colors, extent and locations of illumination, distance from the casting object to the receiving surface, and surface topology of the receiving surface. Here we cover several secondary aspects of information that are perhaps less obvious or familiar.

2.1 Visible Shadow Bodies

Shadows provide information about themselves: a cast shadow has location and shape. But cast and attached shadows also inform us about the volume of (possibly empty) space inaccessible to the light. We call this area the shadow body (fig. 2.1), in other words, the three-dimensional volume that is screened from light (in chap. 7, we will see some twists of the notion of shadow body).

A visible shadow body explicitly links the cast shadow to the object that owns it.

When the air is hazy, foggy, or filled with rain or snow, it is possible to directly see at least part of these boundaries which then appear as light bodies streaming through the atmosphere, accompanied by neighboring shadow bodies (fig. 2.2). When only one object blocks the light, we see an isolated shadow body, as with large buildings or contrails. In figure 2.3, we see that the contrail of a plane that was flying roughly toward the sun creates a thick enough shadow body to filter out light in a portion of the atmosphere. The visible shadow body provides an explicit link between the object and the shadow it casts.

Figure 2.1
A shadow body is the volume of space between the shadow-casting object and the cast shadow where the source of light is not visible. A visible shadow body links the object explicitly to the shadow it casts. Image credit: PC, Hanover, New Hampshire, April 27, 2017.

Figure 2.2
Light bodies are intertwined with shadow bodies. Image credit: Tom Murphy.

Figure 2.3
A shadow body generated by a contrail. The plane is flying approximately in the direction of the sun. Image credit: RC, Saint-Tropez, July 2008.

2.2 The Relative Slant of Surfaces

Shape information is embedded in cast shadows. From features of the cast shadow, we can infer features of the shape of the object casting it. We may start from an extremely basic aspect: if a shadow falling on a flat vertical surface is itself not vertical, then the edge of the casting object is not vertical. In figure 2.4 we can safely assume that the building that receives the shadow and the building that casts the shadow are skewed relative to each other. A generalization of this constraint, holding for flat surfaces, goes as follows:

If a shadow falling on a flat vertical surface is itself not vertical, then the edge of the casting object is not vertical.

2.3 Perspective Information in Shadows

Sometimes, we can assume a shadow in an image to be a self-portrait shadow: the shadow of the person who is viewing the scene (or taking the

Figure 2.4
San Francisco, shadow of the Transamerica Pyramid on a nearby building, late afternoon. Image credit: RC, April 28, 2014.

image). Can we ascertain which shadow is the best candidate for that of the photographer in a populated picture (fig. 2.5)? Can we identify any features common to all self-portrait shadows?

When the sun is behind us, as it is when viewing our own shadow, the antisolar point (sec. 1.2.3) is centered on our head's shadow. So each of the five people in figure 2.5 has their own antisolar point centered on their head shadow seen from their point of view. But the antisolar point for our photographer is the only one to appear in the image. Because of the optical properties of grass lawns (as with evergreen forests and clouds or fog), the reflected light has a peak around the antisolar point directly opposite the sun. Indeed, there is a subtle brightness peak that is centered around one particular shadow in each picture. This peak indicates the antisolar point and identifies the shadow of the "eye" of the camera that takes the self-portrait shadow. No matter how we locate the antisolar point (in chap. 1, we traced the shadow rays), once we have it, we know the observer's or camera's location, and we can use this information to trace back from other shadow features to the scene features that cast them, even when they are outside the frame.

Figure 2.5
Who took each of the pictures? Image credit: RC, Dagstuhl Shadow Seminar, May 2015.

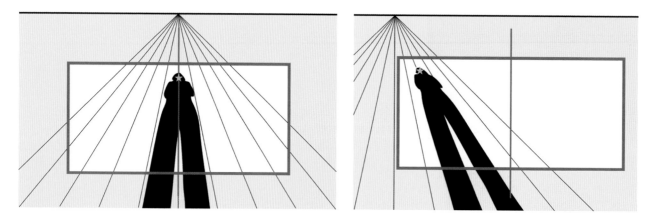

Figure 2.6
Left: No matter where you direct your gaze, your shadow originates from your feet, so if the sun is behind you, it must intersect a line descending vertically from the center of the image, assuming you are positioning the camera vertically above your feet. Here the camera image is the clear central rectangle, and the thinner red line indicates the midline of the image. The antisolar point is indicated by the yellow star in the shadow of the head. *Right*: With the sun not directly behind, the shadow falls along the line between the projection of the antisolar point to the horizon (to which all shadows converge) and the photographer's feet. In both panels, the perspective lines are linear, but they curve to become parallel as they approach the photographer's feet. Image credit: PC.

The orientation of our shadow in an image can also help identify whether it is ours (or the photographer's). Consider standing on a flat surface on a sunny day. Shadows cast by reasonably vertical objects (trees, posts, other people) will appear to converge to a vanishing point opposite the position of the sun. This vanishing point is the projection of the antisolar point to the horizon along the direction of the sun's rays. When our gaze is aligned with the direction of the light, our shadow will be seen as vertical, originating from our own feet and stretching toward the vanishing point (fig. 2.6, left). However, when we (or the camera) gaze at an angle to the illuminant, we no longer see our own shadow as vertical. Even so, we can still identify it as our own shadow by tracing it back to our feet. In particular, the central axis of the photographer's shadow will be the only one to intersect the midline of the frame (fig. 2.6) below the frame, assuming the photographer is standing vertically and holding the camera to his or her eye. The axes of the other shadows will intersect the midline within or above the frame, far from the person's feet.

If a photographer is standing and taking a picture that includes his or her own shadow, that shadow has the antisolar point at the eye or camera, and that shadow's axis intersects the midline of the image.

We can apply these properties to settle controversies about the presence of self-portrait shadows in a painting. For example, Victor Stoichita has suggested that the shadows at the bottom of Renoir's *Pont des Arts, Paris* include the painter's own shadow (fig. 2.7). In contrast, William Sharpe claimed that the painter is situated under a bridge, over which stroll the passersby who cast the shadows.

The geometric analysis suggests that Sharpe is right. The antisolar point is found at the intersection of the red lines, representing the sun rays that connect objects with the shadows they cast (see sec. 1.2.3). Remarkably, the

Figure 2.7
Pierre-Auguste Renoir (1841–1919), *The Pont des Arts, Paris*, 1867–1868. Oil on canvas. Image credit: The Norton Simon Foundation.

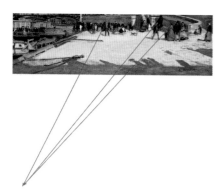

Where the eye of Renoir's shadow should be.

sample rays shown in the margin here do converge to a point appropriately on the left for the sun on the right and behind him, suggesting that Renoir was paying careful attention to the details of people's shadows, even as they moved. The antisolar point locates the shadow of the head of the painter in the scene, here obviously outside the frame. Shadows at the bottom of the painting are therefore from passersby above him, on the then Pont des Saints-Pères (now reconstructed as the Pont du Carrousel).

2.4 Shadows Reveal the Shape of the Light Source

Penumbral width informs us about the size of the light source. For example, the sheer presence of a penumbra reveals that the light source is extended. But can we obtain information about its shape?

Consider figure 2.8. How can the opaque tip of the pen cast a blurred shadow while its edges do not? When we first noticed this phenomenon, we were surprised and tried in vain to move the tip of the pen closer to the surface so as to create a more focused shadow. It took us a while to figure out what was going on. The photograph still looks manipulated to us.

What is happening here? If the light source is linearly extended (like a fluorescent tube) and the object casting the shadow is aligned with it, the resulting shadow will display an exaggerated penumbra at its tip—and only at its tip. If you rotate the object so that its main direction is orthogonal to the main direction of the extended light source, the pen's shadow is all penumbra and weak to the point of invisibility (fig. 2.9). This sensitivity to the light source's orientation is a kind of *shadow anisotropy*. Shadow anisotropy reveals not only the size and extent but also the shape of the light source.

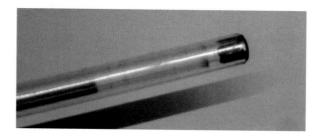

Figure 2.8
Pen shadow, sharp along the pen but blurry at the end. Image credit: RC, MSH Dijon, conference room, April 7, 2015.

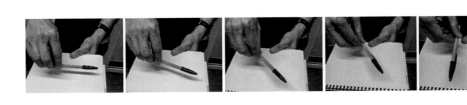 

Figure 2.9
Disappearing shadow. As the pen rotates relative to the neon, its shadow gets so blurred that it is unnoticeable. Image credit: Francesca Bizzarri, Dagstuhl Shadow Seminar, May 2015.

The effect is made dramatic when one of the parts of the object casting the shadow is aligned with the linear light source, while other parts are orthogonal to it (fig. 2.10). In this case, the shadow "loses parts" and may not appear to correspond to the object.

Discrepancies between an object's silhouette and its shadow can reveal the shape of the light source.

Anisotropy can be detected by noticing missing parts, blurred ends, or shadow disappearance upon rotation of the object casting the shadow.

Linearly arranged lightbulbs approximate linear light sources such as fluorescent tubes. As each lightbulb is responsible for a different shadow, the mechanism of shadow anisotropy becomes visually evident.

Shadow anisotropy is a recent addition to the ecology of shadows, as it depends on a particular, artificial type of illuminant. The surprise we may experience when viewing anisotropic shadows indicates that the visual system is not equipped for retrieving the shape of light sources. Artificial vision systems and software for scene analysis, on the other hand, could be designed to take advantage of this feature.

2.5 Movement from Shadows

When an object rotates relative to the light source, attached shadows on the object change, and the shadows it casts move as well. Normally many other cues can reveal the object's motion, although this is complicated when the observer and the object are rotating together.

Figure 2.10
An impossible shadow? One part of the cross casts a shadow, but the other does not. Note the blurred end of the tip of the visible shadow. Image credit: RC, Dagstuhl Shadow Seminar, May 2015.

Figure 2.11
Shadow anisotropy in the presence of linearly arranged light sources. The solid shadow of the pen turns out to be the sum of the shadows cast by each light. Image credit: RC.

Figure 2.12
A roll of paper towel moves through dappled sunlight. The flow of light and shade on the object's surface reveals whether the object is moving toward us or away from us. Image credit: RC, September 2017.

Top: The boat moves relative to the sun-earth direction, exemplified by the shadow. Movie credit: RC, Elba, Italy, July 2015. *Bottom*: The earth rotates relative to the sun-earth direction. Movie credit: RC, Etna, New Hampshire, March 2014. (You can watch both original movies by opening the QR codes with your smartphone.)

For example, an observer on a boat or in a plane may notice movement of the shadows that reveals the movement of the boat or plane relative to the sun. The same change in shadows can be produced if the object were still and the light source moved. The ambiguity about whether the source is rotating around the object or the object is rotating relative to a fixed source has a long history in early astronomy. The motion of shadows on the ground under sunlight has long been interpreted as the consequence of the movement of the sun in the sky, as opposed to the Earth's spin.

Information about an object's movement can be gathered not only from the shadow it casts but also from shadows cast onto the object. If a convex object moves in the dappled shade of a canopy, the pattern of shadows on its surface shows a characteristic motion flow. In particular, if the object is moving toward us, we should see shadows moving upward across the front face of the approaching object. If it is moving away from us, we should see shadows moving downward (fig. 2.12). The principle for retrieving movement of a convex object is thus (assuming light from above) as follows:

If shadows are moving upward on an object, the object is approaching. If the shadows move downward, the object is receding.

2.6 Orientation Ambiguities in Cast Shadows and Silhouettes

Cast shadows are silhouettes of the object shape seen from the location of the light and projected onto a surface. Shadows and silhouettes do not capture the internal features of the objects—features that would be visible within the occluding boundary. They represent only the occluding boundary and necessarily omit all the details within the object.

Silhouettes are ambiguous as to the orientation of front versus back, as both these orientations necessarily share the same silhouette (fig. 2.13). The same ambiguity affects cast shadows. Does the man in figure 2.14 have two left hands (or two right hands)?

Figure 2.13

Left: Is this a left hand (from the back) or a right hand (from the front)? *Right*: Is the hunter shooting toward us or away from us? Image credits: RC, PC.

Figure 2.14

Does the shadow lie? A man with two left hands? A man with two right hands? Image credit: RC, New York, December 2013.

Shadows reveal the object's silhouette from the viewpoint of the light, but like silhouettes, they are ambiguous concerning the front–back orientation of the object.

Point of view plays a key role in disambiguating cast shadows. For example, two interpretations are possible for the three-dimensional appearance of the shadow in figure 2.15. We can see it as if we were viewing the object from above (top sketch on the right) or from below (bottom sketch). A preference for the "from below" interpretation can arise from the fact that the observer is situated at a lower level than the shadow. But the correct interpretation is the "from above" interpretation, as the light source here is the sun, which "sees" the architectural structure from above. The top figure thus delivers the viewpoint of the sun. The limited 3-D information that is typically conveyed by shadows is here overcome by the rich 3-D structure of the object casting the shadow.

Shadows provide information about an object from the viewpoint of the light source, and if the shadow is ambiguous concerning the object's front–back orientation, there can be an even deeper ambiguity about the location of the light source. In particular, we may not know if a shadow is projected on a wall from "our" side or a translucent screen from the "opposite side," which could further reverse our judgment about front versus back.

A viewer's interpretation of the orientation of an object that casts a shadow may be influenced by knowing the position of the light source and of canonical positions of objects. The shadow of a shopping cart (fig. 2.16) could be seen as an image of the cart viewed from below or, more likely, viewed from above. In the former case, the shadow-casting cart would need to be lying on its side, which is unlikely. (Here penumbral differences assist disambiguation: the wire mesh at the top of the image is more blurred than the one at the bottom; hence the corresponding part of the cart is farther away from the ground.)

2.7 Shadow Modulation of Image Detail

A shadow has reduced illumination compared to its surroundings, so the availability of details within the shadowed area depends on the other sources of light that fall within it. This is typically called ambient light and is the sum of all other illuminants, such as reflections of the main illuminant off nearby surfaces, sky light if outdoors (predominantly bluish on a sunny day, as we saw), or other artificial sources. As long as there is

Figure 2.15
The correct interpretation is the view from above (*bottom left*): the way the sun would see the lamp. Image credit: RC, Paris, March 2017.

Figure 2.16
The shadows of 3-D wire objects provide compelling information about their structure but are ambiguous as to the viewpoint. Image credit: RC, Hanover, March 2014.

some ambient light, the details within the shadowed area can be registered, but with any physical imaging device there are limits: too dark, and the details may be lost in sensor noise. Conversely, if the direct illumination falling outside the shadow is too bright, details in the lighted area may be lost in sensor saturation. This effect is often seen in photographs that are overexposed in the illuminated area but well registered in the shadow area (fig. 2.17). Sometimes this effect is also captured in paintings. In Turner's *Cologne: The Arrival of a Packet-Boat, Evening* (fig. 2.18), for example, one notices that the shadow cast on the tower reveals the underlying structure (the stone wall), which is not so clearly visible in the area above that receives direct sunlight.

This is a case in which shadows operate on the information flow, making it easier to process: the extra shadow in our example brings the contrast of detail in another shadow within the contrast range of the sensing device. We may talk of *information enhancement*, similar to the effect you can get by using a magnifying glass or a mirror, or, conversely, of *information loss*, when the details are suppressed.

Figure 2.17
The pole casts two shadows, one from direct sunlight, and the other from a reflection of sunlight in the glass pane. The second shadow, pointing toward the camera, is visible only where the area it is cast on is already shaded. The larger shadow modulates the visibility of the smaller shadow and acts as an information enhancer. Image credit: RC, Boston Logan International Airport, June 2014.

A shadow can reveal details that are overexposed outside the shadow or suppress those that are underexposed within the shadow.

2.8 Summary

We have seen a few of the secondary aspects of information carried by shadows: the volume of space contained in a shadow; the shadow orientation on a vertical surface as a cue for object orientation; the perspective information in shadows that reveal the location of the viewer's or camera's viewpoint; the anisotropy of shadows from linear sources; the cue that shadow motion gives about the motion of the object a shadow falls on. We also saw that the information in the shadows can be ambiguous as to the front–back orientation, and we noted that a shadow can bring the light within it into

Figure 2.18
Joseph Mallord William Turner (1775–1851), *Cologne: The Arrival of a Packet-Boat, Evening* (detail), 1826. Oil and possibly watercolor on canvas (lined), 66 3/8 × 88 1/4 in. (168.6 × 224.2 cm). Henry Clay Frick Bequest, accession number 1914.1.119. Copyright the Frick Collection.

a usable range or reduce it below visibility. Here is the short list of the principles and regularities we have discussed:

A visible shadow body explicitly links the cast shadow to the object that owns it.

If a shadow falling on a flat vertical surface is itself not vertical, then the edge of the casting object is not vertical.

If a photographer is standing and taking a picture that includes his or her own shadow, that shadow has the antisolar point at the eye or camera, and that shadow's axis intersects the midline of the image.

Discrepancies between an object's silhouette and its shadow can reveal the shape of the light source.

If shadows are moving upward on an object, the object is approaching. If the shadows move downward, the object is receding.

Shadows reveal the object's silhouette from the viewpoint of the light, but like silhouettes, they are ambiguous concerning the front–back orientation of the object.

A shadow can reveal details that are overexposed outside the shadow or suppress those that are underexposed within the shadow.

Together with the list of principles described in chapter 1, the environmental regularities we note here offer a rich source of criteria for the visual system in its process of reconstructing a scene. However, being a possible source and being used by the visual system are two different things. Some principles may be ignored, others used only inefficiently. No matter what, to exploit any of the regularities of shadows to resolve the spatial layout of the scene (see chap. 4), the brain must first be able to label the relevant areas of the visual scene as shadows, as opposed to stable dark patches (chap. 6), and then link those shadows to the objects that cast them (chap. 5). But before dealing with these stages in processing shadows, we will lay out the basics of the architecture of the visual system.

The visual system first measures image data, determining color, orientation, size, and other features. From these measurements, it then infers the most likely surfaces, objects, lighting, and shadow. In the image here we see only light and dark regions, and the dark regions could be dark pigment or dark shadow. Many of the contours are characteristic of shoes and laces, so an interpretation of shoes is supported, and much of the dark region is taken as shadow generated by lighting from above and to the left. Image credit: PC.

3   The Visual System

In this chapter, we present a brief overview of the organization of our visual system, with the goal of putting the analysis of shadows in a larger context. (For readers interested in more detail, we list a number of more extensive presentations in the references.)

The vision system is responsible for creating a representation of the scene in front of us. Our experience with cameras may make us think that this is an easy task: just open our eyes and take a picture; and indeed, the initial parts of the visual system, the adjustable pupil, the focusing lens, and the light-sensitive retina in the back of the eye, do act in some ways like a camera. However, cameras and pictures just collect patterns of light, dark, and color. They do not know what those patterns represent (although some recent cameras can now locate some faces). Seeing and recognizing objects and their layout with the accuracy and speed that we have is enormously complicated. One indication of the level of complexity is that visual processing takes up as much as 30 percent of our cortex, the most valuable real estate in our brain. This far exceeds the area dedicated to hearing or language or action. Clearly, seeing has been of extraordinary importance to our survival, and brains have evolved to make vision fast and powerful. Our overview looks at the signals registered by the visual system and what happens to them, with some focus on the processing of shadows. We will see that much remains for us to understand about how shadows are processed and that, in turn—and this is the topic of the rest of the book—shadows offer a powerful window for examining the algorithms and strategies used by vision to understand scene properties.

We divide visual processing into two stages: *measurement* and *inference*. The first stage collects the features of the image: these are measurements of edges, colors, motion, and depth, similar to the data that an advanced

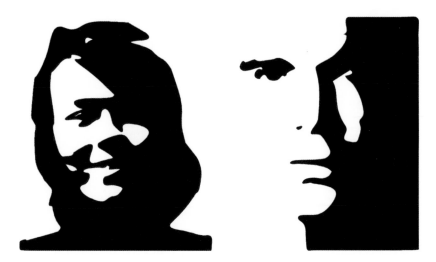

Figure 3.1

Inference in vision. These high-contrast images contain little information, and yet they connect to visual object knowledge; for many of us, this recovers the possible shape of a face. Measurement of the image by a computer vision system can easily find the edges but cannot infer how to piece them together. Our visual system, on the other hand, is very good at doing this. Image credits: PC.

digital video camera might acquire. The measurement stage has been the focus of much of the research in vision and has led to great advances, identifying specific cells in the brain and areas that perform particular parts of the measurement. But measurement alone is not enough. The drawback to pure measurement is obvious when we consider the many ambiguities in each image (e.g., fig. 3.1). Every pattern may have many interpretations: dark areas may be shadows or dark paint or stains; light areas may be reflections, light paint, or light sources. To make sense of these conflicting possibilities and select a final choice of what we see, the visual system adds extremely smart inference processes (fig. 3.2), calling on its own proprietary database of knowledge about objects.

Cognition has its own, separate database of object knowledge and expectations and has little if any influence on these visual inferences. Indeed, cognition and vision are at times in conflict over the interpretation of a scene (typically the case for perceptual illusions), and we discuss this issue further in chapter 7. In many ways, the visual inference processes do most of the complicated work of vision, but we know very little about how this

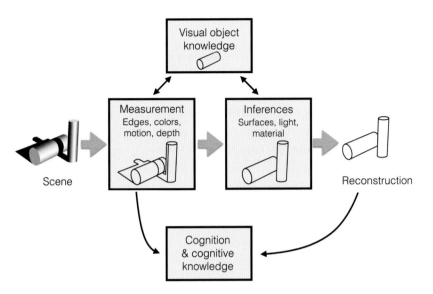

Figure 3.2
In this simplified schematic, visual processing begins with measurements of local features like orientation and color, followed by an inference stage that constructs the best story to explain these measurements. Some familiar objects may be recognized directly from the pattern of measurements; others are recognized only once the bits and pieces are put together in the inference stage. In both stages, visual object knowledge may bias measurements and inferences. The output of the inference stage is a reconstruction of the scene populated by recognized objects, surfaces, light, and shadow. Cognition may reason about information at the measurement or reconstruction stages (chap. 7) based on an independent database of object knowledge but cognition is assumed to be unable to influence visual processing—this is referred to as *modularity* or *cognitive impenetrability*. Image credit: PC.

happens. We understand that it must happen, and we can name parts of what is probably happening, but we cannot point to equivalent cells or regions in the brain that actually accomplish these processes. Recognizing shadows and using the information they provide definitely fall in this second level of inference process. We suggest that a better understanding of shadow information and processing can help advance research on the mechanisms underlying inference mechanisms in general.

We first cover the early visual stages that measure edges and colors, and then we focus on the processes of inference where interpretation, including that of shadows, happens.

3.1 Measurement

In the visual system, measurement involves much more than simply registering the amount of light at each point as a camera would. The image falling on the retina (fig. 3.3) is immediately analyzed for local patterns of light by cells that perform the first in a long series of analyses. Each cell at the retina and beyond is hooked up to a local region of photoreceptors called its receptive field, and the cell computes some property of the pattern of light that falls in its receptive field (fig. 3.4).

At the retina, this is pretty much limited to a comparison of light in the center and surround of its region of analysis to report whether or not there is a local peak of light or local dip. If the light is uniform across the receptive field, the cell does not respond. These comparisons that accentuate change keep the information sent to high levels to an efficient minimum. The signals are then sent via the optic nerve to the next stages of the visual system, where they participate in further contrasts that recover edge orientation, differences in color, and motion. At higher levels, the receptive-field approach continues with larger and larger regions of analysis and many more specialized measurements: motion flow, depth from several cues, and texture, to name a few. The billions of cells in the visual system proceed

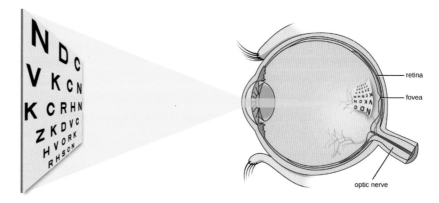

Figure 3.3

The cornea and lens in the front of the eye focus the image of the scene onto the back, the light-sensitive retina. Like a camera, the image is upside-down and left–right reversed. Unlike a camera, the density of photoreceptors changes with location, being extremely dense in the center of vision and much sparser in the periphery. Image credit: PC, redrawn from CNX OpenStax, OSC Microbio 21 01 eyeball.

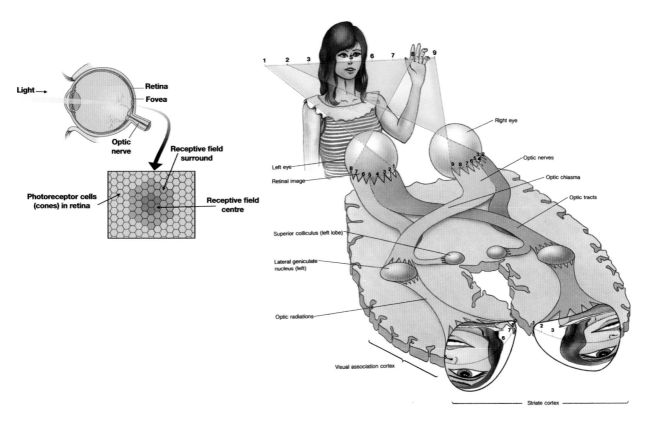

Figure 3.4

Left: Each cell in the visual system is characterized by its receptive field, a region of the retina over which it analyzes patterns of activity to determine if its preferred stimulus is present. Here for a horizontal cell, outputs from a small central group are added together, and the activity in a surrounding ring of photoreceptors is subtracted to generate the receptive field for the cell. *Right*: Information from the eye is laid out in cortical areas like a crude copy of what was on the retina (although not in color, as shown here), greatly distorted with much more area given to analyzing the center of vision than the periphery, and also split in half with the left part of the brain receiving the right half of the world and vice versa. Image credits: Left, PC with eyeball redrawn from OSC Microbio 21 01 eyeball from CNX OpenStax. Right: Frisby (1980).

with a division of labor, each monitoring different areas of the visual scene broken into a large-scale mosaic of receptive fields and registering particular features in their regions of interest. At higher levels, these measurement values for different features may be channeled into separate cortical areas (fig. 3.5). As many as thirty different areas make up the visual system. These areas analyze aspects like color, motion, depth, faces, visual attention, and eye movement guidance and are found throughout all parietal, frontal, and temporal lobes, as well as the visually specialized occipital lobe of the brain.

One indication of the sophistication of receptive-field analyses comes from electrodes recording the responses of cells in the temporal lobes of epilepsy patients being monitored for seizure activity before surgery. Here we see evidence of cells that respond preferentially to individual faces in a variety of poses and settings. In one patient, one cell responded to Jennifer Aniston, while in another patient, a cell had a strong preference for Bill Clinton. The implication is that we all have highly tuned cells or groups of cells responsive to individual faces that are very familiar to us—friends, family, and celebrities.

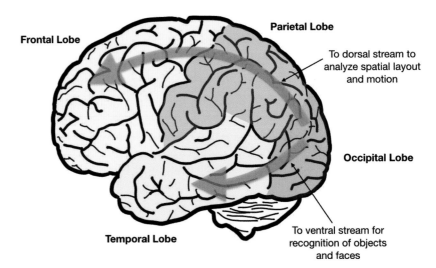

Figure 3.5
The visual areas of the cortex start at the back of the brain in the occipital lobe. From there, information travels up to the parietal and frontal lobes for analysis of spatial layout and motion, and downward to the temporal lobes for analysis of form and face and object recognition. Image credit: RC and PC.

Research on the receptive fields has formed the solid foundation of vision research. To date, the most influential discoveries in vision and the major part of current work can be described as characterizing this measurement component of vision. It is accessible with single cell recordings, animal research, and human behavior. It is understandable that this accessibility has led to impressive discoveries and successful research programs.

3.2 Inferences

Inference is a general label for what the visual system must do next, after it collects the measurements. Processing strategies attempt to make sense of the measurements, making deductions about what is connected to what, what belongs together, what is light and what is a dark surface material as opposed to a shadow. The mechanisms of inference are not yet well understood. Based on how effortlessly we see complex scenes, it is easy to propose that certain inferences must have been made, but our understanding of the processes is still rudimentary. To put this idea in context, we have fifty years of physiological data that show how the measurements of edges, color, contrast, motion, and depth are made. But for inferences, we have virtually no such evidence (with one exception: border ownership). We are really still at the stage of the original proposals of Ibn al-Haytham from one thousand years ago. He outlined a theory of "unconscious inference" that he called the "visual sentient," where intelligent deductions were made based on visual measurements. Hermann von Helmholtz proposed a similar theory in 1867, and since then, vision scientists have made computer models that actually compute some inferences based on simple proposals of how they would work. But we still have little evidence for how inference works in the visual system. Indeed, much of the evidence comes from outside the framework of formal visual research, from the work of visual artists who have been testing the rules of visual inference for centuries (see chap. 10). In this chapter, we cover four different sets of inferences that comprise much of the processing at this level. Inferences must recover object and scene properties like surfaces, illumination, and material properties from the basic measurements. There are many examples where object and event knowledge plays a role in choosing one interpretation over another, so at the end of the chapter, we will look at object recognition and, finally, event understanding.

3.2.1 Surface Completion

The first step in piecing together the parts of an object is to put together its contours and surfaces, a process called *completion* if there is only partial information in the image. In a cluttered scene (fig. 3.6), for example, many objects occlude others, so that we only see parts of them.

But this occlusion leaves clues, because the contours of the farther object make T-junctions as they pass behind the contours of the front object (fig. 3.7). Even if parts of the front object have the same lightness as the background, elsewhere, where the front object is visible, the junctions it makes with the objects behind it reveal its presence to our visual system (fig. 3.8).

Ken Nakayama and colleagues underlined the importance of attributing ownership to a contour: it belongs to the closer surface. Thus the horizontal border in figure 3.7 is owned by the surface in front. If the border were jagged instead of straight, then the front surface would own the jaggedness.

Figure 3.6

Three principles of completion. In normal scenes, objects often occlude each other, so that we see only partial views. But they do not look partial. Our visual system infers that one object has hidden parts of the other, and our internal representations are completed with what we would expect to be there. *Left*: The arm of the front figure hides part of the back of the other figure, making a T-junction where the two contours meet. This provides an informative cue for the continuation of the farther object behind the nearer one, and we fill in based on our knowledge of the form of the human body. We do not see the completed information, but we would be surprised if, from a different viewpoint, we saw a gap in the back figure's body that just happened to match the shape of the arm. *Right*: The T-junctions between the railing and the boy's body indicate that the railing continues behind him. Finally, the water we see through and above the railing all has the same texture and color, supporting the continuation of the same surface behind the railing. Image credits: Left: Caeli Cavanagh. Right: PC.

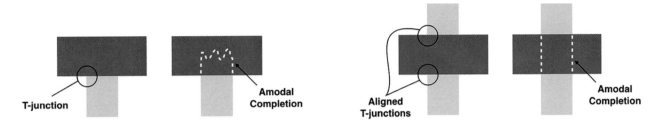

Figure 3.7
Left: A light gray surface meets a darker surface, making T-junctions where the vertical contour abuts the horizontal. We have the impression that the light surface continues behind the dark, although to an uncertain end. This is called *amodal completion*, "amodal" because is it registered but not seen. *Right*: When two T-junctions align, we have even stronger evidence that the light surface passes behind the dark, so that the two light rectangles are actually the same surface. They are linked together by this inference of completion. Image credit: PC.

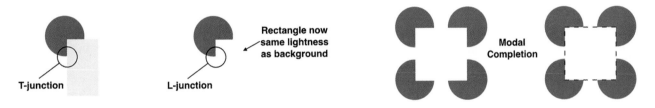

Figure 3.8
Subjective contours. The T-junction becomes an L-junction when the front surface has the same lightness as the background, taking a chunk out of the disk on the left here but producing no impression of the now invisible overlying rectangle. However, when several of these L-junctions align, the invisible front surface now becomes visible as the simplest explanation of all the missing pieces. This is called *modal completion* because it is seen in front, as indicated by the dashed outline on the far right. Image credit: PC.

Pieces of contour of the surface behind it can link up underneath the front surface. Rudy von der Heydt and his colleagues added important physiological evidence to this aspect of border ownership, showing that some neurons in an early area of visual cortex (V2, where V1 is the first area in visual cortex) responded to a line only if it was owned by the surface to its, say, left, whereas other neurons would respond to the same line only if it belonged to the surface on the right (fig. 3.9). This is one of the most impressive pieces of physiological evidence for inferential processes.

The visual choices for how the surfaces are combined are not always logical or compatible with cognitive world knowledge—a horse may come out impossibly long, for example—but these choices of visual inference appear

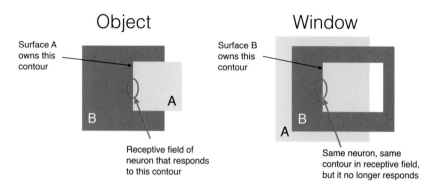

Border ownership. On the left, the front surface, *A*, owns the border; on the right, surface *B* owns the border. The red ellipse indicates the receptive field of a neuron in area V2, like several recorded by von der Heydt and colleagues. They found some neurons that were tuned to orientation were also selective for which surface, the left or right one, "owns" the border. Here a neuron that likes vertical borders responds when the surface that owns the border is to its right (as in the Object example), but not when it is to its left (as in the Window example), even though the local contour is identical in both cases. Image credit: PC.

Figure 3.9

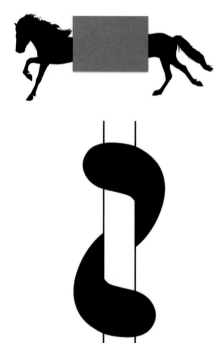

The black form wraps around behind the vertical bar even though it has no aligned image contours on the two sides of the bar.

to be driven by the priority given to connecting collinear segments that both end in T-junctions.

In contrast, an X-junction suggests that the surface is still visible on either side of the junction; it may be due to a transparent surface or a shadow, a reflection, or simply a line or cord lying across the surface (fig. 3.10).

Given this highly lawful behavior, we might ask whether there is anything really inferential here. Indeed, Stephen Grossberg and his colleagues have modeled the majority of these examples within a neural network that requires no appeal to "object knowledge," visual or otherwise. However, these straightforward examples give a restricted picture of the range of completion phenomena.

Peter Tse has shown that there is quite good completion seen even for objects that have no collinear line segments (the black worm in the margin here). Clearly more is going on than can be explained by image-based rules. According to Tse, visual object knowledge can be as minimal as the property of *being* an object—having a bounded volume—and does not require that it be a recognizable, familiar object.

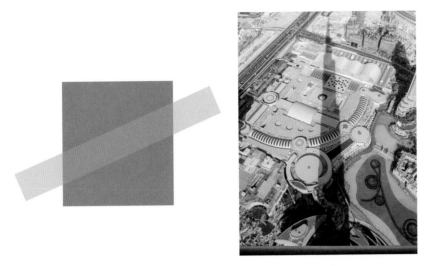

Figure 3.10
X-junctions indicate that neither surface occludes the other. *Left*: A transparent front surface crosses another, and the borders cross over. *Right*: The shadow of the Burj Khalifa crosses over ground contours. Image credits: PC.

3.2.2 Illumination: Light and Shadow

Each region in a scene sends some amount of light to the eye, and the task of the inference stages is to explain the intensity of this light. If little light arrives, it may be a shadow area blocked from direct light, or a surface at a sharp angle to the light, or simply a dark surface material in full illumination; if a lot of light arrives at the eye from a particular location, it may be a light source itself, or a reflection of a light source (a highlight), or a well-lit white surface. To decompose the received light into its possible factors of surface reflectance (pigment), orientation to the light, and illumination, we must make many assumptions about the scene and the objects in it. Usually the final choice is a compromise that involves the fewest assumptions and breaks the fewest rules. For example, the dark area in figure 3.11 follows many of the principles for shadows that we outline in chapter 6, and is seen as a shadow even though it does break at least one rule: it has the wrong shape.

If the choice is that a region is a shadow, it is seen not as a dark surface material but as a region of low illumination. The reflectance seen in its surfaces is corrected for the reduced illumination (fig. 3.12). The light reaching

Figure 3.11
A shadow region is taken as a change of illumination, not a change in pigment. The observer makes the inferences of light and reflectance even though the area seen as a shadow is not the correct shape. Image credit: Nisha Sandhu.

the eye from a surface is the product of the light reaching the surface and the reflectance of the surface. To recover the reflectance, the interesting, intrinsic property of the material, we need to discount the illumination— meaning that somehow we must recover what that illumination is.

Gilchrist has suggested a solution whereby the visual system divides the visual scene into regions of reasonably constant illumination that he called *frameworks* (and Adelson called *atmospheres*). Gilchrist further suggested that the visual system assumes that the brightest surface within each

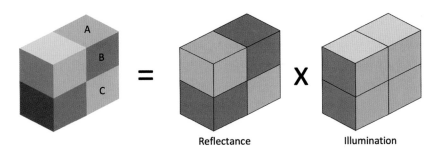

Reflectance   Illumination

Figure 3.12

Lightness correction. The pattern of light arriving at the eye is the product of the reflectance of the surface and the illumination that falls on it. Here the illumination is from the top and to the right, so that the top surfaces of the blocks receive the most light. Our perception discounts this variation in lighting in an attempt to recover the actual reflectances of the surfaces. As a result, *A* and *B* on the left are seen to be the same gray surface, whereas *C* is lighter than *B* and *A*, even though *C* and *A* have equal luminance (i.e., send the same amount of light to the eye). Image credit: PC, adapted from Adelson.

framework is taken as white, with a reflectance, therefore, of 1. So this surface acts as an anchor and reveals what the illumination is for that region. The reflectances of other surfaces in that framework are then found from their ratios relative to the anchoring patch. Gilchrist has many compelling demonstrations that something like this in fact holds; we do in fact see the brightest surface in a region, if there are enough patches of varying lightness, as white, even if it is not.

Gilchrist also showed that our judgment of the illumination depends on where the probe appears to lie in a scene—either in one framework or in the other. To do so, he built adjoining rooms, one brightly lit in back and one dimly lit in front, with a door between them. In one condition, he placed a test patch so that it appeared to lie on the far wall of the well-lit back room, whereas in the second condition, it appeared to lie in front of the door frame in the near, dimly lit room (fig. 3.13). Both cases used the same patch (always attached at the door frame), with the same luminance. Gilchrist just added notches in the shape to make it appear to lie behind the two adjacent cards in the first condition. Subjects reported that the test looked almost black in the far room, apparently under bright light, but almost white in the front room under dim light. The point is that when

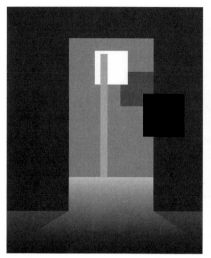

Figure 3.13

Position in the scene affects which illuminant will be discounted. *Left*: Two rooms are visible, the front one with dim illumination and, seen through a door, a back room with bright illumination. The middle test card, which is always fixed at the same location on the door frame, is made to appear to be in the brightly lit back room using only pictorial cues (the notches next to the adjacent cards). *Right*: The arrangement is identical except that without the notches, the middle test card now appears to be in the dimly lit front room. Although the card always has the same luminance, subjects see it as a dark gray material when it appears to be in the back room (*left*) but a light gray material when it appears to be in the dim front room (*right*). (The effect in this drawing is much less pronounced than in the actual three-dimensional experimental setting.) Image credit: PC, adapted from Gilchrist.

the visual system discounts the illuminant, the location of the object in the scene can determine which one to discount.

Although the visual system may be able to parse a scene into regions of different illumination, there appear to be limits to how complex these regions may be. For example, if we see a small region of different illumination from its surround, we may assume the region belongs to the larger framework, resulting in large errors in apparent reflectance. These effects were first explored by Gelb, and one example is shown in figure 3.14, where a light source has a hole in it that is aligned with a white disk that appears to be a deep black. The border between bright and dim illumination falls on the object border, and there is no independent evidence of an illumination

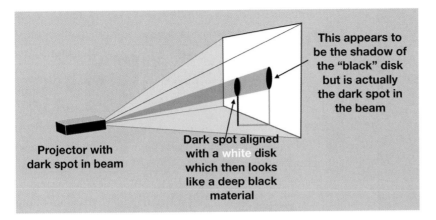

Figure 3.14
When a light source has a hole in it that aligns with a white disk, the disk appears to be black. The illumination from the light falling on the background is assumed to fall on the disk as well, so to account for the low luminance, the observer sees the disk's surface as black. Image credit: PC.

change, so none is assumed, a situation we explore again in chapter 6. If the disk is now moved partly into the light, the shadow border becomes visible, and the disk is seen to be white with the adjacent dark section seen as shadow. If the disk is realigned with the dark spot in the light, the disk immediately appears black again, though the observers know it is white. The inferences made by the visual system have no memory even though the cognitive system does.

The important lesson is that once a shadow has been identified as such, it also provides information about spatial layout (chap. 4). The separation between the object and its shadow influences the object's perceived three-dimensional location in the scene. The processes linking the shadow and the object (chap. 5) are, however, quite tolerant of discrepancies (recall fig. 3.11) that are physically impossible. Little is known about the physiological underpinnings of shadow processing. Research on patients with brain damage has suggested that inferences about shadows are processed within the temporal and parietal lobes.

3.2.3 Reflections

The task of determining if a region is a shadow has a complementary task of determining whether a region is a reflection (see chap. 9). A smooth

 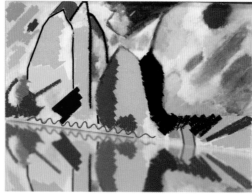 

Figure 3.15
Left: Reflections on horizontal surfaces show vertical symmetries. *Middle*: artists use the vertical symmetry to depict water even in a highly abstract composition. *Right*: Surface highlights are like shadows but must be lighter than their surround, as they are an addition of light. Image credits: Left, Don Graham, *Tuolumne Pond Reflections, Yosemite NP 5-15*. Middle: PC redrawn from Kandinsky, Autumn II. Right: PC.

reflecting surface copies surrounding details that are visible in the reflection, and this has characteristic symmetries. Horizontal reflecting surfaces like water have a vertical symmetry between the original and the reflection (fig. 3.15), whereas the symmetries in vertical mirrors are horizontal. Few animals understand reflections on vertical mirrors, but they do understand horizontal reflections (from the surface of water). Local specular reflections, like highlights on a surface, are the inverse of shadows in that they must be lighter than the surround, and the highlight contours must make X-junctions with surface contours that it crosses.

3.2.4 Material Properties
Inferences are necessary as well to determine what kinds of surface are in a scene.

Are they glossy, matte, transparent, rough, smooth, wood, water, jam, or toast? Each of these has characteristic local features of texture and pattern that bias the interpretation in its favor. Transparent materials are interesting in the context of shadows because they share a critical defining feature: the borders and patterns on transparent surfaces must make X-junctions

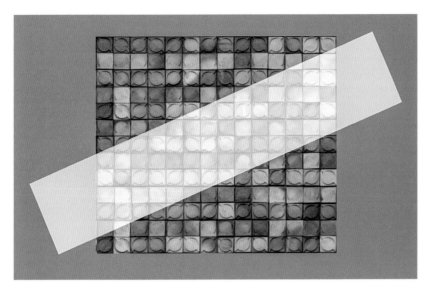

Figure 3.16
Transparent materials make X-junctions with underlying contours, and the textures and properties of the material under the transparent cover should match the textures and properties just outside it. Image credit: PC.

with the surface contours they cover (see chap. 9). The texture of the surface we see through the transparent material should match the texture seen directly adjacent to the transparent material (fig. 3.16). Artists use these cues widely to depict transparent materials.

3.3 Objects

The goal of the inference process is to link the bits and pieces of the image together so that we can recognize them as objects that have some spatial arrangement in the scene. In many ways, the process is a circular one, because, as described earlier, some of the inference steps are guided by visual object knowledge that has been triggered by a quick recognition of a probable object based on a distinctive cue or a familiar contour. Thus, when we see a new object, we attempt to identify it from a few cues and then impose our internal model of the expected object on the image data. For human bodies, we expect a head, torso, arms, and legs, with a particular top-to-bottom and front-to-back orientation. These axes may be ambiguous

Figure 3.17
Our visual system interprets objects based on distinctive features and expected structures. *Left*: The shoes bias us to see these front limbs as legs, although the resulting
interpretation of the expected body parts connected to those legs then fails. *Right*:
The head on the sand seems to belong to the torso with the hidden head. The bias to
complete the body schema overrules the unlikely separation of the body parts. Image
credits: Left: from Cavanagh (2011). Right: from Cavanagh (2011).

or unspecified in the image, but we assign them anyway, sometimes arriving at unexpected organizations (fig. 3.17). These unexpected percepts,
going far beyond what is specified, offer evidence for the internal axes and
part structure that our visual system insists on applying.

3.4   Events

There is more to vision than just recognizing objects in static scenes. The
true power of vision is its ability to predict, to see things coming before
they happen to you. And the most useful information for prediction is the
motion of objects in the scene. In fact, it is so useful that two separate
motion systems appear to have evolved quite independently, one a reflex-
ive, low-level system and the other an active, attention-based, high-level

system. The low-level system does not call on inference or other advanced processing strategies, but the high-level system does. Motion can tell us more than where an object is going; it can also tell us what the object is. The characteristic motions of familiar objects, like a pencil bouncing on a table, a butterfly in flight, or a closing door, can support the recognition of these objects. In return, once we recognize the object and its stereotypical motion, knowledge of that motion can support the continuing percept. Like the first notes of a familiar tune, our knowledge can guide our hearing of the remainder of the melody, filling in the missing notes. Point-light walkers provide a compelling example. An observer can easily recognize a human form from the motions of a set of lights attached to a person filmed while walking in the dark.

The idea that there is a story behind a motion percept is a simple version of the intriguing effects of intentionality and causality. The original demonstrations by Michotte for causality and by Heider and Simmel for intentionality have captivated students of vision for decades. These effects demonstrate a level of "explanation" behind the motion paths that is, to say the least, quite rich. It suggests that the unconscious inferences of the visual system may include models of goals of others, as well as some version of the rules of physics.

Beyond the logic, the story, and the intentions implicit in perceived motion lies an entire level of visual representation that is perhaps the most important and least studied of all. Events make up the units of our visual experience like sentences and paragraphs do in written language. We see events with discrete beginnings, central actions, and definite end points. This syntactic structure of the flow of events undoubtedly influences how we experience the components within an event as closely spaced in time just as the Gestalt laws describe how we see grouped items as closer together in space than they are. The phenomenology of motion perception has provided one of the richest sources of examples for the inference stage of vision: motion interpretations can undergo dramatic reorganization under the influence of object knowledge, attention, and instruction. Given our focus here on mostly static images, we will not discuss this highest level of visual storytelling in depth, but it is important to mention it, as it is indeed active in many examples of shadow/surface mark disambiguation: shadows display specific patterns of motion and even more specific patterns of movements of X-junctions.

3.5   What Does Vision Tell the Rest of the Brain?

The descriptions of surfaces, objects, and events computed by visual processes are not solely for consumption in the visual system but live at a level that is appropriate for passing on to other brain centers. Clearly, the description of the visual scene cannot be sent in its entirety, like a picture or movie, to other centers, as that would require each of them to have their own visual system to decode the description. Some highly compressed, annotated, or labeled version must be constructed that can be passed on in a format that other centers—memory, language, planning—can understand. This is the idea of a central bulletin board or chat room where the different parts of the brain post current descriptions and receive requests from each other, like perhaps "Hey, Vision, are there any red things just above the upcoming road intersection?" The nature of this high-level visual description is as yet completely unknown. We can imagine that it might be what we label as conscious vision, if only because consciousness undoubtedly requires activity in many areas of the brain, so visual representations that become conscious are probably those shared outside strictly visual centers.

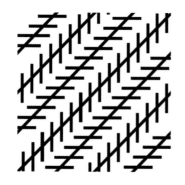

Vision can make extraordinarily sophisticated inferences that are totally separate from standard, everyday, reportable cognition, as indicated in the schema at the opening of this chapter (fig. 3.1). For example, knowing that the oblique lines are all parallel in the Zöllner illusion in the margin does not make them look so (for further discussion of the independence of cognitive knowledge and visual knowledge, see chap. 7). Given that the brain devotes more than 30 percent of the cortex to vision, we can certainly imagine that the "visual brain" is a smart one. What is appealing about this separate visual intelligence is that its mechanisms of inference may be easier to study, unencumbered as they are by the efforts of ordinary cognition. So when we look at what science has uncovered about visual inferences, we of course believe that these processes may be duplicated in other regions of the brain that deal with the broader conscious level of cognition.

3.6   Summary: What Is Vision?

Although researchers have made remarkable progress in understanding the level at which inferences are processed in vision, it is perhaps worthwhile pointing out that many of the major questions were identified much earlier.

They certainly formed the core of Gestalt psychology in the 1920s. These phenomenological discoveries—subjective contours, ambiguous figures, depth reversals, visual constancies, rules for grouping and parsing the visual scene—have filled articles, textbooks, and classroom lectures on perception and philosophy of mind for the last one hundred years. What has changed recently is the degree to which computer programs have been developed to simulate these high-level effects. Work on object structure, executive function (memory and attention), and surface completion has been an active focus of research, although the pace has perhaps slowed recently. In its place, driven by brain-imaging work, many vision labs have focused on localization of function and on the interactions of attention and awareness. But understanding localization and awareness (if that were ever to happen) is not enough to uncover the rich and sophisticated mechanisms of vision.

So what is vision? On the large scale, visual processes construct a workable simulation of the visual world around us, one that is updated in response to new visual data and serves as an efficient problem space in which to answer questions. The representation may encompass the full scene or focus only on the question at hand, computing information on an as-needed basis. This representation forms the basis for interaction with the rest of the brain, exchanging descriptions of events, responding to queries.

3.6.1 How Is It Computed?

What are the computational processes underlying vision? We have hinted at a number of different computational styles, from measurement to inference, from recognizing faces to understanding intentions. A different architecture undoubtedly operates in the measurement stage, with hardwired receptive fields implementing powerful neural networks that passively vote for the presence of one shape, color, motion, or object over others. The inference stage uses much more flexible computations in a majority voting scheme similar to a relaxed constraint satisfaction scheme (fig. 3.18). It begins with an army of heuristics analyzing the local structure of the scene: T-junctions, X-junctions, symmetry, anchoring, distinctive features, and many others that we will introduce in the following chapters. Each of these votes for or against several possible interpretations. The interpretations themselves then access knowledge about how such an interpretation should look: if it may be a car, it should have wheels; if it may be a person,

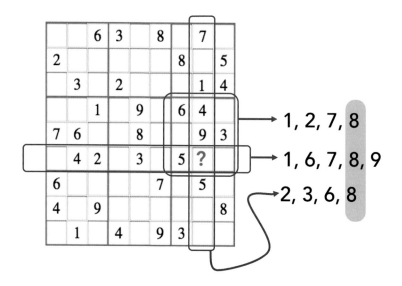

Figure 3.18

An illustration of constraint satisfaction. In sudoku, the rule is that each column, each row, and each of the nine subblocks of nine squares must contain all the digits 1 to 9 with no repetitions. For the cell marked with a question mark, there are then three constraints, none of which on its own specifies a unique answer, as shown to the right. However, only one digit, the 8, satisfies all three. In a similar way, the inference engine for vision considers multiple cues, each voting for possible interpretations, with the difference that only a majority is needed to trigger the choice (thus some cues voting against the choice may be overruled). This flexibility is critical for making rapid decisions. Image credit: PC.

it should have arms and legs. These verification details can then be checked in the image to see which interpretation receives the most support, and can even pass elements into perception that are not present but could be. Higher-level constructs like goals, causality, and animacy may play a role in determining what is expected in the scene. All these levels would ideally work interactively in selecting the most appropriate interpretation; that is, they are not stepped through in an ordered sequence.

The point of this flexible architecture is first of all that it relies on simple cues that can be computed locally and rapidly—we will see this bias to locality when we explore the properties that count in labeling shadows. Second, by using these cues independently in a voting system, it sidesteps the problems of noise that would derail any serial, analytic approach when a single critical cue in a sequence is incorrect. One key consequence of a voting architecture is that when several cues point to one interpretation, some other cues that are trying to veto it may be overruled. The result is that the application of the heuristics or rules or critical properties may seem idiosyncratic: they are important in some contexts but ignored in others. In the chapters that follow, we will refer to this architecture simply as flexible and idiosyncratic and underline in many ways the strengths that emerge from this arrangement.

3.6.2   How Can We Test It?

This description of the architecture of inferences is necessarily vague. We have just described how we think it ought to work. It would be nice and logical, for example, if cues to completion actually trigger choices about surfaces. But as mentioned before, we do not have much evidence that any of this takes place, despite the plausibility of it all. The one physiological example we have of computations at this level is von der Heydt's border ownership cells in area V2 of visual cortex. We need many more break-throughs of this sort before we can have any confidence in specifying the actual architecture. Perhaps there is no inference or choice at all but just powerful mechanisms of neural learning that can learn what things ought to look like based on sparse, noisy cues. One of our main goals in this book is to emphasize the opportunity that shadows provide in the hunt for the mechanisms of inference. Many of the principles we describe for making choices about what is or is not a shadow can lead to new research on the underlying inferences. At the end of each chapter, we highlight these opportunities in a "Further Research Questions" section.

Further Research Questions

1. Have we learned to extract information from new, artificial environments that does not exist in natural settings? Have we developed any sensitivity to unnatural or absent penumbrae, or shadows with unecological shapes (such as linear light sources)?
2. Simple brain-imaging and physiological experiments should be able to show where in the visual cortex shadows are identified and discounted (fig. 3.19) and so move us along in uncovering the mechanisms of visual inference. Far from being well understood, the mechanisms of shadow and object perception are still very much up for grabs.

 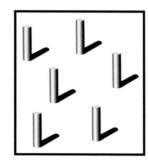 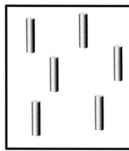

Figure 3.19

A hypothesis: Shadow borders may not have the same status as object borders in the visual system. On the left, receptive fields of cells in early visual cortex may respond equally vigorously for an edge that is a shadow border and an edge that is an object border. However, at higher levels in the visual system, the shadow border may be suppressed, and cells will no longer respond. In the middle, we see a set of posts with shadows, each forming an *L* shape, whereas on the right the same posts have lost their shadows and form a set of *I*s. The activity in the different regions of the brain can be registered with fMRI while subjects are looking at the posts with and without shadows, and an algorithm can be trained to distinguish whether the subject is looking at *L*s or *I*s on the basis of the activity patterns. This classification should be quite successful in early visual cortices. At higher levels, where shadows are suppressed, the *L* shapes will no longer be represented, both patterns will be just *I*s, and the discrimination will fail. Image credit: PC.

3. A shadow provides a second view of the object casting the shadow, approximating what the light source would see if it were provided with an eye. The shadow image is in the norm distorted relative to the view from the light source's viewpoint. How much of the object's shape can be retrieved by an artificial vision system compared to how much can be retrieved by a human observer?

Endeavour's final launch, May 16, 2011. Its trail casts a long shadow on the clouds below. Image credit: https://apod.nasa.gov/apod/ap110525.html.

4   The Shadow Mission

Shadows fill our environment with a wealth of information (chaps. 1 and 2), which is available to visual brains like ours, whose architecture we sketched in chapter 3. But the path to recovering this information is not straightforward. The information in an image is inherently ambiguous, and our visual system weighs alternatives based on assumptions about objects and light and the likelihood of each interpretation. A critical part of the interpretation is labeling some regions as shadows, and we discuss the factors that contribute to this labeling in chapter 6. Some shadows also have owners—the objects that cast the shadows—and we discuss the bases of assigning ownership in chapter 5. Once an area is labeled as a shadow and has been attributed to an owner, its contributions to scene layout and illumination can take place, and that is what we present in this chapter. The order of the chapters may seem to invert the natural progression, but as we mentioned, the solution to what is or is not a shadow, and what object may or may not have cast the shadow, is not straightforward. Indeed, the decision to label a region as a shadow will often depend on the plausibility of the scene information it supports. For that reason, we too begin with the information the shadow brings to the scene: its mission.

We know something about how shadow labeling and ownership and layout interact because it is possible to stress-test vision in ecological and experimental settings, as we shall see in this chapter. Visual artists have been conducting these experiments for millennia to discover how to convey scene layout and lighting using flat surfaces and a limited set of luminances. In particular, artists have widely explored the appearance and use of shadows that deviate from photorealistic accuracy but still act as shadows (we discuss this subject further in chap. 10).

Humans are only one of many species to use shadows for understanding scene layout. A plausible selective advantage that drove the development of vision was the detection of a predator's cast shadow that would pass over the prey species, alerting it to danger. Crayfish, for example, use the movement of a shadow projected on the ground to execute antipredator behaviors. Shadows can also serve to *hide* information in the poor lighting available there; western scrub jays use shaded areas for caching food, in particular when in the presence of other scrub jays. In humans, the sensitivity to shadows to judge depth is not present at birth. By seven months (but not at five months), infants will reach to an object that appears nearer because of its cast shadow. Similar early use of shading for depth judgments is seen in chimpanzees at six to twelve months.

Here we look in more detail at what shadows provide to human understanding of scenes. We describe how vision uses the presence of a shadow to correct for lighting and recover actual material properties in the scene. We demonstrate how cast shadows reveal the presence and locations of objects in the scene and can indicate whether an object rests on a surface or floats above it. Shadows also contribute to the recovery and recognition of shapes, either from the silhouettes of a cast shadow or the surface relief in attached shadow borders. Once we extract this information from shadows, the shadow mission comes to an end. Shadows do not vanish from our view, but they are demoted to lower status in terms of salience, and their borders are removed from consideration as object contours. An inglorious end, perhaps, but we nevertheless enjoy the scene structure that they have helped build before they step aside.

Finally, we also explore conflict situations where shadows tell the visual system one thing, but other evidence, like geometric cues, indicates something different. Shadow interpretations are often overruled in these instances.

4.1   Discounting the Illuminant

When a region is taken as a shadow, it is seen not as a dark surface material but as a region of low illumination. The reflectance seen in its surfaces is corrected for the reduced illumination, and this correction forms a critical part of the shadow mission. It is not necessary to know what cast

The shaded parts of the white tiles are darker than the sunlit parts of the gray tiles, but perception corrects for illumination. Image credit: RC, San Francisco, May 2014.

Figure 4.1

Left: The square of light and dark tiles is crossed by a diagonal alternation of shadow and light. *A* and *B* are seen as light tiles and *C* as a dark tile even though in the image *A* and *C* have the same luminance. Their perceived reflectances have been corrected for the apparent pattern of shadow and light. *Right*: In the absence of a valid shadow border (one that does not match object borders), there is no sense of variation in illumination across the tiles, and we now see both *A* and *C* as medium tiles and *B* as a lighter tile. Image credit: PC.

the shadow to make this correction; simply labeling the region as a shadow suffices.

To demonstrate the correction objectively, we can ask observers to match test points in the image to comparison patches presented separately on a fixed, say, white background. On the left in figure 4.1 is a pattern of shadow and light on a checkerboard where all the tiles are either of a light or dark material. An observer would pick a lighter comparison patch for tile *A* than *C* in this case, and similar comparison patches for *A* and *B*. But *A* has much less light coming from it than *B*, and the decrease has been attributed to a lower level of illumination—tile *A* is in a shadow. *A* and *C* actually send the same amount of light to the eye but do not look like the same material because of the apparent difference in lighting. The strength of the correction for the low light in the shadow is evident on the right in figure 4.1. Here the darker top and bottom corners and the lighter middle swath are the same as on the left, but now the border between the dark and light regions follows exactly the edges of the tiles. The change is therefore assigned to the tile material itself, and now we see three tiles, light, medium, and dark, all with the same illumination. In the right panel of figure 4.1, an observer would pick more or less the same comparison patch for *A* and *C* and a lighter one for *B*.

These tests for discounting the illuminant are direct and objective tests of whether a region is perceived as a shadow and we will use these again in chapters 5 and 6. The outcome of the test on the left versus the right in figure 4.1 indicates that the shadow border on the left has properties that no longer hold on the right. These properties are among those described in chapter 6 and are specifically that the border should cross the underlying surface borders, making X-junctions, and should not be aligned with the surface features.

Shadows indicate areas of reduced illumination, triggering a correction of surface reflectance within the shadow.

Here is a critical caveat that holds true for many of the images that we present. As a reader looking at figure 4.1, your perception may differ somewhat from our description of the typical observer's response, and that is normal given the variations in people's perception. However, our description of the typical observer's response will always be supported by a study that you will find in the bibliographic notes at the end of the book. In the case of figure 4.1, for example, the relevant studies are by Adelson, Gilchrist and colleagues, among others.

4.2   Recovering Depth

If there is one thing cast shadows are really helpful for, it is the understanding of depth in the scene. A cast shadow reveals that there is distance between an object and a surface: that the world is not flat. How good is vision at using this piece of evidence?

The scene layout we see in figure 4.2 appears to match pretty much what we might expect from a reverse engineering of the physics of light and objects. We can retrieve positions of the objects and see that they are all sitting flush on the ground plane. We call shadows from objects that sit on the surface "anchoring shadows" because their contact with the object indicates that the object contacts the surface on which the shadow falls (more detail in chap. 5). When an object does not lie flush on a surface, the offset of the shadow relative to the object indicates that the object floats over the surface on which its shadow falls (fig. 4.3).

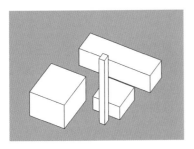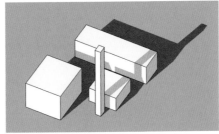

Figure 4.2

Left: Without shadows. Is the elongated horizontal parallelepiped above or beyond the flat box? *Right*: The shadows place the objects relative to each other and anchor the objects to the ground plane. Image credit: RC.

Figure 4.3

Anchoring and floating. On the left, the location of the cube without a shadow is ambiguous. The next cube has an anchoring shadow that meets its base, indicating that the cube sits on the ground plane. The shadows of the next two cubes are detached from the cube's base, indicating that they float over the surface. The last two on the right also show that shadows accomplish their mission even if their shape is not correct. Image credit: PC.

Finally, two-tone images of faces (fig. 4.4) give a good example of recovering surface relief from the pattern of cast and attached shadows, and we explore this phenomenon in more detail in section 4.3.3.

Shadows contribute to spatial layout.

When the direction of illumination is not known or not recoverable, human observers appear to assume a fixed lighting direction. This is seen here for static objects (fig. 4.5) and mentioned again later for moving objects (fig. 4.7), and it makes a change in shadow offset appear as a change in depth, when it could equally well have been a change in the position of the light source. This is another example of the idiosyncratic logic for shadows:

Figure 4.4

Left: A two-tone image of a woman's face. The depth relies on the shadow structure to reveal facial relief. *Right*: It is clear that this depends on shadows because when the contrast is reversed, the shadows and depth are lost. Image credits: PC.

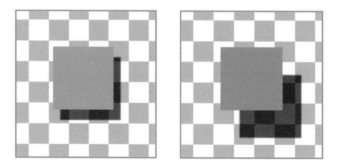

Figure 4.5

The impression of a separation between the squares and the checkered background increases when the shadow is offset farther from its square. Although this shift in shadow position could have been caused by a change in the position of the light source, the interpretation strongly favors a change in depth over a change in the light source. Image credits: PC.

Recovery of depth from cast shadows assumes a fixed light source.

Note that a further assumption is at work here: the fixed light source is assumed to be not very near the observer (fig. 4.6). Consequently, small shadow offsets that could be due to a light source near the observer's location are equated with small object-to-surface separations, generating illusions of closeness. We would have no need for this assumption if the human visual system could recover the illumination direction as computer vision systems do (e.g., from multiple shadows, highlights, shading), but this feat seems beyond our capabilities, perhaps because it is too computationally demanding or not often necessary. This bias that a *small offset equals a close surface* ignores possible differences in illumination direction. Still, it is a good-enough heuristic to navigate a nonflat world.

A small offset between an object and its shadow is interpreted as a small separation.

Figure 4.6
The effect of the "small shadow offsets imply small separations" bias can be seen in flash photography where the flash is close to the camera and its cast shadows have only a slight offset from the objects that cast them. This leads to an impression of compressed depth. Here the white paper seems quite close to the background wall although it is about 1.5 m away. Image credits: RC.

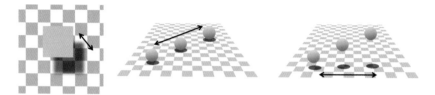

Figure 4.7
If the cast shadow of the square on the left slides back and forth, it gives the impression that the square has moved toward and away from us in depth; the motion of the shadow creates a motion perception for the square even though it has no image motion. Similarly, if a ball moves along an oblique path, the 3-D location of the ball is ambiguous, but the position of its shadow can specify a path along the ground plane (*middle*) or rising in the front plane (*right*). Image credits: PC, based on Kersten, Mamassian, and Knill (1994).

Scan these codes for videos of the stimuli at the Kersten lab demo page. http://gandalf.psych.umn.edu/users/kersten/kersten-lab/demos/shadows.html.

As a special case of scene layout from shadow, vision uses shadows to reconstruct movement in the scene. Whenever an object moves, so does its cast shadow. This can even produce an impression of object motion when the object has little or no motion itself (when the object moves along the line of sight), or resolve the three-dimensional trajectory of an object's motion when it would be otherwise ambiguous (fig. 4.7). These inferences of object motion from the motion of their shadows assume that the light source is fixed; this seems to be a basic assumption made by the visual system. Scenes in which the light source does move often lead to difficulties in interpreting spatial layout.

We will use this ability of shadows to provide relative depth as probes to test what is a shadow and what is not a shadow in chapters 5 and 6. These tests for shadowness show that the "shadow" logic used by our visual system is, as described in chapter 3, rough and idiosyncratic in several ways. Evolution proceeded by trial and error to find partial or just good enough solutions for using shadow information, consolidated them, and passed them on from ancestral visual brains to our contemporary visual system. We should not expect that the solutions include all the relevant information, and we should expect that the solutions perhaps overgeneralize to include somewhat irrelevant information. We see this in the use of cast shadows to recover separation between an object and the surface on which the shadow falls. As the cube in figure 4.3 on the right showed, the separation is recovered even when the cast shadow does not match the appropriate shape. The degree of mismatch that can be tolerated is still an open research question. How inappropriate can a shadow be and still accomplish its mission? In chapter 5, we will see that a coarse match suffices to assign ownership and lead to an estimate of spacing. Tolerance for incongruent shadows sits well with the idea that shadows' representations are mainly position indicators. Reflections also serve this purpose (chap. 9) and, very much like shadows, indicate position, and a mismatch between the reflected shape and the original object also goes mostly unnoticed.

4.3   Recovering Object Shape

4.3.1   Cast Shadows

In general, human observers can often recognize complex solid objects from their cast shadows (see chap. 7). Nevertheless, this is only crucial when the object casting the shadow is not visible or its shape is not recognizable in the direct view (fig. 4.8).

Figure 4.8
A cast shadow can reveal the object's shape as a silhouette. See chapter 7 for more on recognizable shadows. Image credits: Left: Kingshield Windows. Right: RC, Barbizon, August 2017.

The statue of the rooster is hard to recognize in this top view.

Light rays (*white arrows*) are perpendicular to the surface normal (*black arrows*) at the terminator. Image credit: RC.

In these instances, the shadows are simply acting as a second view of the object and are recognized as independent entities. Some cognitive juggling may then transfer the identity to the source object. We might think that cast shadows would interfere with recognition because of the noise that the shadow borders inject into efforts to segment objects from each other and from the background. Although research offers some evidence for interference, one study using images of fruits and vegetables with and without shadows found that the time and accuracy in naming objects was unaffected by the presence of shadows.

4.3.2 Attached Shadows and Shading

Attached shadows are regions of an object that are blocked from the main illuminant. The attached shadow contains no shape-specific information. The critical shape information from the attached shadow comes from its boundary, the attached shadow border, which traces the line on the object where the light just grazes the surface (the illuminant is at 90° to the surface normal). This suffices for a rich shape recovery even in two-tone images (see section 4.3.3). However, much of the research interest in attached shadows is related to the ambiguity that they produce in surface recovery. This was first pointed out in the second century CE by Ptolemy, who noted that the curvature of a distant sail catching the sunlight would flip between convex and concave. The light and dark patterns on the surface relief can be consistent with concave or convex shapes *depending on where the light is coming from*. Here again the visual system makes an assumption about light position (we have already seen that it takes the light position as fixed and not near the observer). It turns out that where no other cues disambiguate the lighting direction, we have a strong bias to assume the light is from above (fig. 4.9).

Recovery of depth from attached shadows and shading assumes that light comes from above.

What does "from above" mean? The convex–concave inversion is experienced within a limited set of lighting directions, centered on a direction that is close to straight up. Moreover, "above" relates not to gravity but to the orientation of the head. If the implied light source is above us in the world when we are upright, then it is to the left in the world when we tilt our head leftward and under us when our head is upside down. Looking at

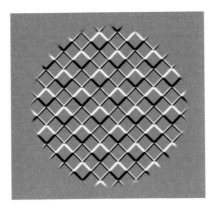 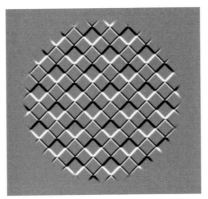

Figure 4.9
The same pattern of light and dark is interpreted in two very different ways depending on orientation. Try turning the page upside down, and then try turning your head upside down instead. Image credit: Pascal Mamassian.

When lit from below, the craters are seen as bumps. "Snoopy" lunar module, May 1969. Image credit: NASA.

figure 4.9 with our head upside down reverses which picture we see with the individual convex tiles.

The light-from-above (i.e., from the direction of the top of our head) assumption finds a possible explanation in the environmental statistics. In the vast majority of ecological situations, the light comes from the sky, and we are standing right side up—light comes from above us. A study with chickens raised from birth with light from below (from the enclosure floor) still showed a light bias from above, so the bias may have been imposed over an evolutionary timescale.

The assumption is resistant to cognitive knowledge. Consider the "bumps" on the moon seen at the right from the Apollo 11 lunar orbiter. You may know that no such formations exist on the moon, but you still see them as convex bumps. The only way to invert the concavity–convexity perception is to turn the page upside down—or to turn your head upside down. Resistance to cognitive penetration provides evidence for an independent early labeling of shadow and light by the visual system that cannot be influenced by our cognitive expectations.

In some real-life situations, one may experience concavity–convexity inversions, based on the default assumption that light comes from above, together with the peculiar shape of the shadow. For example, the "truncated pyramid" of figure 4.11 is actually a recess in a wall.

Figure 4.10

The left-hand image shows pieces of wood lying on a workbench. When the same image is flipped upside down (*right*), it appears as a convex cutout of a stag, an interpretation that forces itself on us in spite of being wrong. Image credit: Walter Wick.

Figure 4.11

Illusory pyramid. Image credit: RC, Paris, 2012.

4.3.3 Shape Recovery from Shadows in Two-Tone Images

Shadows help us recognize the objects that cast them if they project a familiar, canonical silhouette, but shadows can also reveal the surface structure of an object to help understand its shape directly. In this case, it is the attached shadows and the self-shadows (cast shadows from one part of the object onto another) that allow this recovery. The most dramatic example of this shape recovery comes from two-tone images where the image has been thresholded into two levels of brightness (fig. 4.12). Two-tone images are often referred to as Mooney figures after Craig Mooney, who used them in perceptual tests in 1957. Earlier examples can be found in the art of Giorgio Kienerk (fig. 4.12, left).

The thresholding of the image makes areas that are dark because of dark material and areas that are dark because of shadows all the same, making the recovery especially challenging. Nevertheless, this technique works quite well, at least for familiar objects, and despite its sparseness, it gives vivid, three-dimensional impressions (fig. 4.12). It is so effective that

Figure 4.12
In two-tone images, the dark areas can be dark pigment or dark shadow, and the light areas can be light pigment or light reflections. Nevertheless, the images support a dramatic 3-D recovery. They present a challenge for recognition because we cannot identify the dark shadows based on image properties: they only become possible shadows in the context of the object, but the object cannot be recognized unless the shadow areas are labeled. We describe a solution based on contour matching in the text. Image credits: Left: Giorgio Kienerk, *Lev Tolsto'i*, as published in the journal *Avanti della Domenica*, September 18, 1904. Center, right: PC.

high-contrast images have become a staple of graphic art. (We will use these images in chap. 6 as an effective test for shadow labeling. For instance, we will see that the depth in two-tone images is destroyed if the shadow region is not darker along the border or if another contour is aligned with the border.)

Two-tone structures do not occur in natural scenes but are nevertheless readily recognized by infants and newborns. This suggests that the objects are recovered not by specialized processes that have been acquired to deal specifically with two-tone images, which newborns are unlikely to have encountered, but by general-purpose visual processes capable of disentangling dark shadow and dark pigment based on visual object knowledge. These processes would have evolved for the natural environment, where redundant cues often help to dissociate dark shadow from dark pigment. In the case of two-tone images, however, only visual object knowledge is capable of guiding the extraction of shadowed objects. Thus two-tone images are useful tools that can give us access to these shadow logic processes in isolation.

How is the object's shape recovered? One possible mechanism is based on accessing stored three-dimensional shapes of familiar objects that have characteristic contours (fig. 4.13). The two-tone image of the face on the top left is easily recognized, but its contours (top center) are less understandable—they could be taken as a map of, say, Central America. There is nothing that links the little loops of contour together. However, among those contours is a set of characteristic face contours, shown on the left, which are the attached shadow edges that conform to particularly informative ridges of high curvature like the nose, eyebrows, and lips. In addition, though, there is another set of edges from the self-shadows cast by parts of the face onto other parts. These edges are just accidents of light and are not informative, but so far there is nothing to distinguish the informative attached shadow contours from uninformative ones that come from cast shadows. To break this impasse, we must assume a process that can match image contours to those of a prototypical face while ignoring the unmatched noise contours. If a match is found, this suggests a candidate object that might explain some of these contours. Once that candidate object—here a face—is selected, it can then be checked against the image: "Is there anything that can explain the unmatched contours? Shadows?" And if the shadow explanation is supported, then the stored

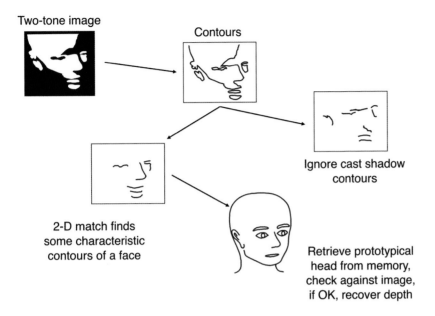

Two-tone image

Contours

Ignore cast shadow contours

2-D match finds some characteristic contours of a face

Retrieve prototypical head from memory, check against image, if OK, recover depth

Figure 4.13
How do we recognize two-tone images? The isolated contours for this face are un-recognizable on their own, but characteristic face contours are embedded along with the uninformative cast shadows. If a matching process can detect these characteristic contours in the presence of the noise contours and find support for a shadow expla-nation (e.g., if adjacent areas are appropriately dark), then stored structural informa-tion about faces can be used to recover the 3-D shape. Image credit: PC.

knowledge about the three-dimensional structure of faces can be used to generate the three-dimensional impression that we experience. Of course, if no appropriate explanation for the unmatched contours was forthcoming (e.g., the image was in negative contrast, so the putative shadow areas were not appropriately dark), then we would have to reject the face hypothesis and consider an alternative.

The account makes a critical assumption that a matching process can check the full contour set against all possible candidates, quickly, perhaps within one or two tenths of a second, the common unit of visual recogni-tion processing. This already suggests a powerful parallel memory process, but it also implies that we can recognize these two-tone images only if they are of familiar objects, like faces, cars, boots, some tools, and so on, for which we have stored prototypes. In fact, researchers have run this test, and indeed, only familiar objects give the vivid 3-D impressions in the two-tone version. In figure 4.14, the same lumpy parts have been arranged on the left to make a face a bit like a Buddha sculpture, whereas on the right the parts make an amorphous jumble. In the thresholded versions, the Buddha is easily recognized as a face with clear 3-D structure, whereas the unfamiliar shape looks more like a 2-D cutout, perhaps of a pirate ship; the actual 3-D shape is not recovered.

The suggested account may or may not hold up, but no process for recognizing two-tone images has been found to work unless it incorporates knowledge about the possible objects to be recognized. In sum:

Knowledge about an object is necessary to recover its shape from shadows in a two-tone image.

For our purposes here, the exact mechanism of recognition is not as important as the use of these images to test whether shadows are acceptable or not (see chap. 6). Object knowledge can only guide shape recovery from a two-tone image if the putative shadow areas meet the requirements for shadows. For example, when the contrast of a two-tone image is reversed, shadow areas are lighter than their surrounds, violating one of the basic shadow principles (again, see chap. 6). As a consequence, the depth from shadows is lost and the shape they would support is often unrecognizable. The perception of depth in a two-tone image is therefore a sensitive test for shadow principles that can or cannot be violated.

4.4 Demotion, the End of the Shadow Mission

After we label a region as a shadow and extract useful information from the scene, what then happens to the shadow regions? They are not objects themselves but merely accidents of light, and the presence of their contours constitutes an endless source of noise for perceiving and interacting with the objects in the scene. Indeed, the visual system had to evolve to recognize objects in arbitrary lighting and so must be able to ignore cast shadows. This has led to a view that cast shadows are "expendable," not especially useful after their information is extracted, and potentially distracting. Note that once demoted, shadows do not vanish; they just become less salient. We can process them when asked, and reason about them (see chap. 7).

Figure 4.14
Both objects in each pair are made from the same ellipsoidal parts, combined in different ways. *Left*: The parts are combined to make a face: its two-tone version is easily recognized and has a rich 3-D structure that is a reasonable match to that of the original grayscale version below. *Right*: The parts are combined in an unfamiliar, lumpy jumble, and its two-tone version conveys little or no 3-D structure that matches the original grayscale version, appearing instead more as a 2-D cutout. Image credit: PC and Cassandra Moore.

Example of shadow noise (competition with visual objects). Image credit: RC, Ithaca, Greece, July 30, 2017.

A number of visual search studies have supported this argument for the discounting or demotion of shadows. In one, several shapes are presented, each casting a shadow (fig. 4.15, left), and in half of the trials, one of the shadows is misaligned with the others. The task is to report whether the misaligned shadow is present. The search for the odd shadow in this case is relatively slow. In the control condition (fig. 4.15, right), the cast shadows are inverted in polarity, light on the gray background rather than dark. Now the item with the odd orientation is easier to find, presumably because the lighter areas are not treated as shadows and thus are not discounted. The hypothesis is that the early visual system rapidly labels regions as shadows, extracts relevant scene information, and then discounts them, making their shapes and contours less of a nuisance in analyzing the scene and attending to the objects of interest. As a result, the properties of the shadows themselves are more difficult to access. Note that the odd direction of the target shadow did not veto it as a shadow; otherwise the shadow would be easy to find. Further studies have shown that finding an inconsistently shadowed item in a complex scene turns out to be easier than in the simpler scenes used here.

Importantly, in the tests shown in figure 4.15, it is specifically the property of the shadow, such as its shape or orientation, that is the target of the search. In this case, search is *slow*, as the shadow has been demoted. In other visual search studies, it is the depth or scene property produced by

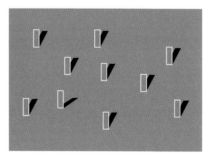

Figure 4.15
The odd dark shadow (*left*) is harder to find than the geometrically identical odd lighter tab (*right*). The assumption is that on the left, the dark areas are labeled as shadows and demoted as features of light rather than features of objects. Since they are less salient, they are harder to find. Image credit: PC, adapted from Rensink and Cavanagh (2004).

the shadow that is the search target, not the shadow itself. In that case, the search is *faster* when the shadows are appropriate, as they produce robust shape or depth features that are easy to find, though the shadows themselves are less noticeable.

Shadows have reduced salience.

In chapter 6, we use this visual search technique to determine which properties are critical for determining if a region is or is not a shadow.

4.5   Shadow Conflicts

Scene layout can be recovered from a multitude of cues including perspective, occlusion, binocular disparity, motion, and, of course, shadows. What happens when shadows indicate one layout, but other cues suggest a different one? Mostly, shadows lose. For example, geometric cues to three-dimensional shape easily overpower shadow-based interpretations. Here we look at some battles that shadows fight, winning some, losing others.

4.5.1   Geometric Cues versus Shadow

Mach cards are pieces of stiff paper that have three folds (fig. 4.16), and when viewed with one eye closed to eliminate cues from binocular disparity, they have ambiguous depth structure. They may be seen as an M shape with the middle fold going back in depth, or a W shape with the middle fold coming forward. With some mental effort, most observers can alternate between these two interpretations. The consequences of these reversals on the shading, shadow, and material interpretations of the paper are striking.

Once the incorrect interpretation is established, the shadows on the Mach card's flaps are no longer consistent with the light source, and the paper may seem strangely transilluminated on the lighter strips and dark in the shadowed strips. Even more dramatic is the reinterpretation seen when a second object, here a pen, is placed across the card and the depth assignment flipped. Now the cast-shadow cues to the separation of the pen from the paper are simply ignored, indicating that geometric cues dominate shadow cues in determining depth. The rule is here robust:

Geometric cues overrule shadow cues.

We see a similar dominance of geometric cues and "good" form over shadow cues in figure 4.17.

Figure 4.16

The Mach card has two depth interpretations when viewed with one eye closed: the actual depth and a reverse depth where the convex folds become concave. (This requires some mental effort: imagine the card is hanging from a vertical wall, for instance.) When the depth is reversed, material and shadow cues to the appropriate 3-D structure are ignored, and perspective information determines the perceived shape. Image credit: PC, based on Mamassian, Knill, and Kersten (1998).

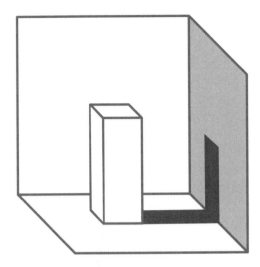

Figure 4.17
Cube or internal face of a corner? The cast shadow should disambiguate the shape, but you can see the background figure as a convex cube, and the cube interpretation is resilient despite the impossibility of the shadow on the right. Image credit: RC.

4.5.2   Face versus Shadow

Another case of resistance to the disambiguating power of shadows is that of the hollow face. When we look at a mask from the inside, the surface of the face should look like the concave relief that it actually is. The shading and shadows clearly support this interpretation, and yet the surface is invariably seen as convex. Many suggest that a lifetime's experience with faces that are convex, as they should be, makes us perceive the face with its more familiar natural depth structure, inverting the apparent direction of the illumination at the same time. Here again the depth from shadows is overruled by other factors, now the expected convexity of faces. The hollow-face illusion also resists other disambiguating signals such as binocular vision and motion: as we move around a stationary mask, the geometry of the inverted depth requires that the face appears to turn with us, keeping us in view, and we may even experience a paradoxical double rotation (the perceived face spins in one direction, and at the same time the mask frame spins in the opposite direction).

To sum up:

Face knowledge overrules shadow cues.

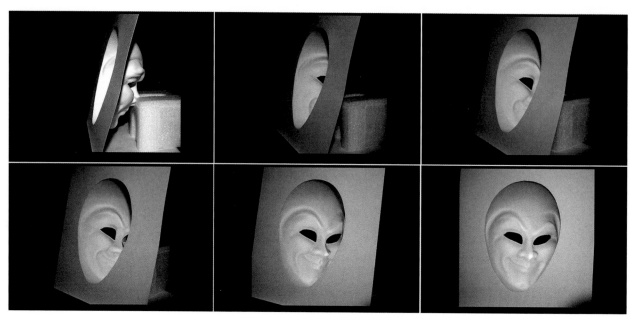

Figure 4.18
What we perceive as a convex face is actually the back of a mask. All the faces here are hollow masks, evident at the beginning of the sequence but not elsewhere. Shadows are reinterpreted by changing the apparent direction of the illumination and do not assist in recovering the correct interpretation. Image credit: Francesco Pia.

We can observe the preference for convexity in other forms that do not easily reverse their depth as the lighting direction changes. Some of these may involve knowledge of familiar shapes, like feet or torsos, but others just stubbornly prefer to be seen as convex (fig. 4.19) in spite of their implausibility (nobody leaves convex footprints). These biases may explain the variability in the ability to mentally reverse footprints or traces of familiar objects. For some people, the concave interpretation is not easily available.

4.5.3 Continuity versus Shadow

Shadows are visual patterns with shape, size, and location. One might expect that they would be subject to principles of perceptual organization that would group together any separated parts into a whole. However, when shadows fall across different surfaces, there appears to be no process that pieces them together. For example, the pen on the left in figure 4.20 appears to have two shadows. The two end shadows appear to join under

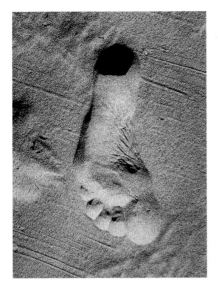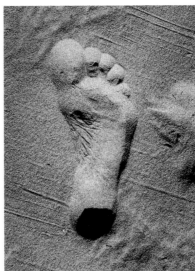

Figure 4.19
A footprint with light from above (*left*) and light from below (*right*). For many, both views appear convex. Image credit: RC.

the book, while the central portion makes a separate shadow on the book itself. On the right in figure 4.20, the faucet's shadow is broken in two parts that seem simply unrelated shadow bits. In contrast, when object parts are misaligned by optics (in the margin), our visual system tries to mentally link them together, revealing processes that attempt to stitch together the various parts of an object into a whole. This does not appear to happen for shadow parts, suggesting that the grouping processes sensibly ignore the nonobject shadow regions.

Shadow shapes are piecemeal.

Separate shadow parts can link together if they are accidentally aligned (fig. 4.21, left). Visual continuity destroys the information about the change in height at the curb.

4.6 Conclusions

We live and evolved in a world full of shadows. Vision learned to discount them when they create noise, but at the same time to take advantage of

Figure 4.20
Left: a pen with two shadows. *Right*: two portions of the shadow of a faucet. Image credit: RC, Barbizon, 2010.

Figure 4.21
In the image at left, the pavement appears continuous; visual continuity masks a conspicuous step, visible in the image at right. Image credit: RC, Paris, September 2015.

their rich information. In this chapter, we saw how shadows underpin the perceived distribution of light and the location of objects in space, following some flexible rules that deliver good approximations. We also witnessed shadow information being easily overruled by other cues.

Shadows indicate areas of reduced illumination, triggering a correction of surface reflectance within the shadow.
Shadows contribute to spatial layout.
Recovery of depth from cast shadows assumes a fixed light source.

A small offset between an object and its shadow is interpreted as a small separation.

Recovery of depth from attached shadows and shading assumes that light comes from above.

Knowledge about an object is necessary to recover its shape from shadows in a two-tone image.

Shadows have reduced salience.

Geometric cues overrule shadow cues.

Face knowledge overrules shadow cues.

Shadow shapes are piecemeal.

But what tools launch shadows on their mission? A quote from Pascal Mamassian provides a road map of things to come. Shadows are loaded with information, he observes, but "to use that information, our visual system has first to segment regions in the image, decide that these regions are potential shadows rather than, say, ink blots, and then match these shadow candidates with objects in the scene." These initial steps are the topics of the next two chapters. Chapter 5 explores how shadows are linked to the objects that cast them, and chapter 6 covers which properties are necessary for a region to be labeled as a shadow and which are not.

Further Research Questions

1. What accounts for the variability in shadow-induced concave–convex reversal? Not everyone seems to be able to reverse the depth.
2. We pitted some factors against each other (shape versus shadow, shadow versus continuity, etc.). A full map of conflicts would be useful. When is shadow a clear winner, when is it a clear loser, and what influences the outcome of the competition?
3. Does shadow agnosia (inability to recognize shadows) exist and would it hinder object recognition?
4. What exactly is the mechanism whereby shadows are discounted? What is the status of the discounted scene information attributed to shadows?

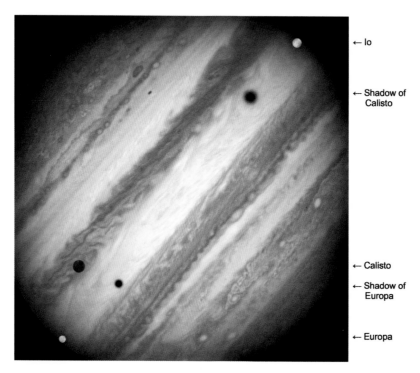

← Io

← Shadow of Calisto

← Calisto

← Shadow of Europa

← Europa

Whose shadow? Three moons of Jupiter are visible in this image: Io (*top*), Callisto (*middle*), and Europa (*bottom left*). Io and Callisto appear to own the shadows that are closest to them, but in fact, these are not their shadows. Shadowless Io at the top has stolen Callisto's shadow, and Callisto has in turn stolen Europa's. January 24, 2015. Image credit: NASA, ESA, and the Hubble Heritage Team (STSc/AURA).

5 Shadow Ownership

5.1 Matching Shadows to the Objects That Cast Them

We have seen how shadows have a mission for vision, and how they accomplish it. The visual system exploits the shadow to recover scene properties that we outlined in chapters 1 and 2: for example, the position of an object relative to its shadow helps gauge the distance between the object and the surface on which the shadow is cast, as well as the position of the light source. But wait: before the visual system can use a shadow for reconstructing the spatial properties of a scene, the visual system must know *whose* shadow it is. Here we will call this the *shadow ownership problem*. Astronomers, for example, cannot use shadows to gauge the position of Jupiter's moons relative to the planet (see the opening image) until they can say which moon owns which shadow. So how do they, and how does the visual system, determine shadow ownership?

Ownership can be seen as an instance of a more general problem. Vision builds complex groups out of individual tokens that share some common property, such as proximity, similarity, or connectedness. It then treats those groups as individuals in their own right (fig. 5.1). This parsing dramatically simplifies the scene: instead of registering twelve or sixteen separate items, vision registers three rows (left, middle) or four columns (right). One further principle is common fate: objects that move together are grouped together. In the early decades of the twentieth century, Gestalt psychology explored the principles that guide this grouping process.

We will see in section 5.2 that the rules for assigning shadow ownership also follow these Gestalt principles. Specifically, the ownership link between an object and its candidate shadow is determined by proximity, connectedness, similarity, and common-fate relations. We will then see

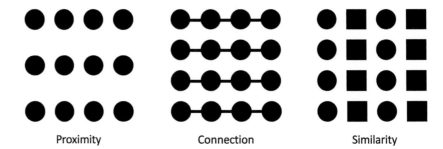

Proximity   Connection   Similarity

Figure 5.1
Gestalt principles of grouping. *Left*: The dots in the rows are closer together than in
the columns and appear to form three horizontal groups. *Middle*: With spacing the
same horizontally and vertically, the connection along the rows favors grouping the
dots as horizontal rows. *Right*: Again with equal spacing, shape similarity favors seeing vertical columns. Image credit: PC.

some unintended consequences of these rules, in particular when it is not
possible to apply them or when their application creates shadow theft: the
dark region may be attributed to the wrong object, and the shadow is captured by an object that does not cast it. In some cases, when an appropriately dark region does not find an owner, it may be rejected as a shadow;
it thus fails to be labeled as a shadow and becomes instead a surface marking. In other cases, it is still labeled as a shadow and just stands alone as a
shadow whose casting object is out sight—a shadow orphan.

5.2   Rules for Shadow Ownership

Let us first consider the world around us. In most ecological situations,
objects are generally quite close to their shadows. Rocks and trees protrude
from the soil, and terrestrial animals move close to the ground. Shadow
ownership is normally a simple matter of proximity. In the case of visual
clutter, however, when many items populate the scene, assigning ownership
for each shadow becomes challenging (fig. 5.2). The ecology of "separated"
or *detached* shadows is also quite rich. Foliage, birds, and clouds provide
interesting instances of objects that are remote from the surface on which
they cast their shadow. And in conditions such as sunrises and sunsets,
even the shadows of terrestrial items are long and distant enough to create
ambiguous or difficult-to-solve images. Celestial shadows (e.g., the earth's

Figure 5.2
Each object has its cast shadow, and each shadow its owner, but it is not always easy to assign individual ownership. Proximity is not very helpful in this case. Image credit: RC, August 2017.

shadow cast on the moon) are so far away and difficult to mentally model that they frequently create ownership ambiguities: eclipses puzzled early sky watchers, and even today many educated adults believe that phases of the moon are the result of a shadow cast by the earth, as if we were permanently in a lunar eclipse.

The solutions to the shadow ownership problem are in many cases idiosyncratic, in the sense that the rules are flexible so as to make the most of the visual scene. We will see in chapter 6 that a shadow does not have to have an owner to be seen as a shadow; furthermore, in some cases, ownership is simply impossible to assign. However, in most cases, a shadow does have to have an owner for it to contribute to the recovery of scene layout. A similar theme of idiosyncratic solutions and flexible rules emerged in chapter 4 in our discussion of the recovery of scene information from shadows.

5.2.1 Shadow Proximity

Pascal Mamassian proposed that shadow ownership is solved at a coarse-scale analysis of the visual scene. The coarse representation ignores the detailed features of a shadow and the object casting the shadow and matches the visual center of mass of the shadow to the object with the nearest center of mass. The advantage of coarse matching is speed. Later refinements are always possible, but if our visual system is to use shadows for retrieving shape or distance, we can reasonably expect that ownership is processed quickly.

The crude-proximity solution for shadow ownership is likely a consequence of the fact that few ecological shadows have features that match them unambiguously to the object casting them: in most cases, shadows of three-dimensional objects on complex surfaces are just messy, amorphous visual blobs that do not share the color, texture, or shape of the objects that cast them (see the image in the margin). A precise assessment of the projection would be computationally expensive. Thus:

A perfectly recognizable object and its poorly diagnostic shadow. Image credit: RC, Paris, August 2017.

Figure 5.3
Whose shadow? If the shadow belongs to the object to its left, both the lighting direction and the distance from the shadow to the object will be different than if the shadow belonged to the object to its right. Note that in this case the visual system initially assigns a shadow by default to the closest object, the one partially covering it. Image credit: RC.

Proximity of a dark region is a strong cue to shadow ownership.

Mamassian's stimuli were limited to simple pairings between an object and its likely shadow and did not include cases of potential conflicts that might arise in more complex scenes (e.g., fig. 5.2). Another issue is the scope of coarseness; are there any limits on the acceptance of whatever dark region might be nearest? We will see that, in fact, some shapes are just not accepted as possible shadow matches.

5.2.2 Shadow Connection

When an object contacts a surface, it will generally cast its shadow on that surface, and that shadow has to meet the object at its base. These are called anchoring shadows (chap. 4). In this case, the connection between the shadow and the object is critical for determining shadow ownership. In chapter 6, we present several examples where a misalignment of the shadow at the object's base vetoes not just the shadow ownership link but also the shadow label. For example, figure 5.4 here shows how a shadow has to connect appropriately to an object that is sitting on or emerging from a surface, and when it does, it becomes an anchoring shadow.

In figure 5.5, breaking the connection to the shadow at the object's base may make the shadow appear orphaned, but still a shadow of something.

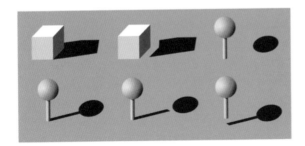

Figure 5.4
For an object that contacts a surface, its cast shadow must align appropriately with the object's base, and all the shadow parts must be connected to this base. The contours of the two shadows on the left mate appropriately with the object contours that cast them. They appear as shadows, owned by the objects they contact. In the four examples on the right, the shadow contours do not connect appropriately with the object. Without any other properties promoting a shadow interpretation, the absence of an ownership link also vetoes the shadow label, and these then appear to be surface markings. Image credit: PC.

Figure 5.5
When the connection is broken between the shadow and the base of the stop sign that casts it, the shadow no longer belongs to the stop sign. Because of other cues, however, it may still appear to be a shadow. Image credit: PC, Outremont, Quebec, 1978 (original photograph and digital modification of the right image).

Similarly, in figure 5.6, the square dark region is contiguous with the post but does not appear to be its shadow. It violates the connection rule: the shadow borders are not tangent to the object's borders at the point where the shadow begins. But though the ownership link to the post is rejected, the area does not lose its shadow label. Because of its other properties, such as an internal grass and gravel texture that matches the external texture, it still looks like a shadow, but of some object outside the field of view.

The ownership effects of connectedness have not been systematically explored with objective tests such as recovery of depth from the offset between the shadow and the object. For the moment, we summarize these effects as follows:

Ownership of anchoring shadows requires a good connection.

5.2.3 Shadow Shape

The cast shadow's shape needs to have only a qualitative similarity to the shape that the object would cast. For anchoring shadows, this tolerance

Figure 5.6
Bad connection: the square shadow is not seen as a shadow of the post. Image credit:
RC, Hanover, New Hampshire, April 2014.

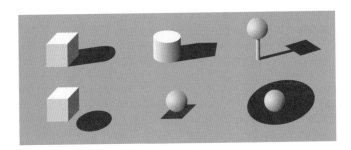

Figure 5.7
In the top row, with anchoring shadows, portions of the shadow distant from the
base can deviate from the expected shape without breaking the ownership link. The
deviation cannot be too extreme, however, and the bottom row shows that detached
and semidetached shadows should not be too different or too large. These mismatches in shape may destroy the shadow labeling as well as ownership. Image credit: PC.

only applies to portions of the shadow away from the connection to the base. Nevertheless, the shadow shape cannot be too different from its expected projection; it needs to be approximately conformal.

The shape of the cast shadow is exactly specified by the location and extent of the light source, by the shape of the casting object, and by the surface relief on which it falls (fig. 5.8, left). But this is a complicated relationship, requiring impressive computational power to work out in computer graphics. The human visual system appears only to take the time and effort to check a few elements of the object-to-shadow shape match. The tolerance for deviation appears to depend on the separation of the shadow and the object that casts it: the three types of cast shadows first described in chapter 1—anchoring shadows, semidetached shadows, and detached shadows—each impose different rules. In addition, if the object casting the

Figure 5.8

Left: Surface relief determines shadow shape. *Right*: The shadow is from an arch far to the left, outside the field of view, making it impossible to check if the shadow's shape corresponds to the projection of the object's shape. Image credits: Left: RC, Paris, November 2015. Right: PC, Jerusalem, January 1985.

shadow is not visible or not linked to the shadow, there are no shape constraints in the first place (fig. 5.8, right).

Anchoring shadows. When an object sits on a surface and casts a shadow on that surface, the shadow serves to anchor the object to the surface, so that we see it as flush with the surface. In this case, the shadow contours must align appropriately with the object contours, beginning at the object contours that cast the shadow. The essential properties for the anchoring shadow are only shown with a few examples here (fig. 5.9) and are only demonstrated by the subjective impression that the dark areas appear or do not appear to be shadows (this is the shadow character, which we discuss in chap. 6). The actual properties remain to be determined, and this will require objective tests of shadow disruption (e.g., loss of depth from shadow or loss of correction for illumination).

Semidetached shadows. When an object is separated from the surface on which its shadow falls, the object may occlude part of its shadow, a situation we call a "semidetached" shadow (fig. 5.10). In this case, the critical properties for the shape match are less restrictive, but present nonetheless. The complexities of the cast shadow shape and the effects of diffuse light

Figure 5.9
Anchoring shadows. When an object sits on a surface, its shadow borders need to align appropriately with the object contours that cast them. This is achieved in the top two examples, and both shadows appear shadowlike, even though the top portions of the one on the right do not have the appropriate shape. In contrast, the bottom two violate this principle and do not look subjectively like shadows. Image credit: PC.

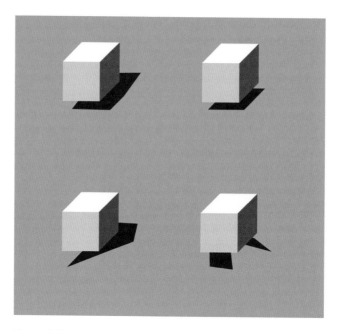

Figure 5.10

Semidetached shadows make T-junctions with the objects that partially occlude them. They make the object appear to float, and relatively imprecise shape properties are tolerated. The top left shadow conforms to the appropriate physical shape, but the top right shadow is too small. The appearance of floating is perhaps even stronger in the top right, suggesting that the T-junctions contribute to the effect. However, the dark regions in the bottom two examples also have T-junctions with the object but appear more as surface markings than shadows. Image credit: PC.

make the physical shadow shape quite variable, so the tolerance to shape mismatch makes ecological and computational sense.

Detached shadows. This is the classic case of a remote cast shadow that is nevertheless owned by the object that casts it (fig. 5.11). The effect of floating that a detached shadow produces is strong in natural scenes like that of the boat in the margin but sometimes ambiguous in simplified diagrams like figure 5.4. The bottom left example in figure 5.11 is more effective where the added blur is consistent with a diffuse light source that would make the cast shadow smaller than the object casting it. The blur alone is not enough, though, as the right-hand example shows, where the shape mismatch is too extreme (although still ambiguous for some observers). The necessary shape properties for an acceptable shadow have

Figure 5.11
Detached shadows do not contact the object that casts them in the image or in the scene. The floating that an accurate, sharp shadow should depict is perhaps ambiguous in the simplified diagram on the top left, and the dark region may appear instead as a surface marking with the cube flush on the surface rather than floating, as it also may on the top right, where it is much smaller. If the region boundary is blurred, as in the bottom left, the floating and therefore the shadow labeling are more effective. But arbitrary blurred dark shapes like those on the bottom right may lose the shadow labeling for some observers and again look like surface markings. Image credit: PC.

not yet been systematically explored. For the moment, we summarize this as follows:

Shadow ownership requires that the shadow shape be approximately conformal.

When the shape mismatch is too extreme, the ownership link may be broken or absent (but still leave the dark region looking like a shadow, once more a topic for chap. 6). For example, the shrub in figure 5.12 is actually casting the elongated shadow to its right, generated by extremely tangential light at the point of contact of the leaves with the wall, but the shadow seems too dissimilar to belong to the shrub. It is still seen as a shadow, though, just not of the adjacent leaf that actually casts it.

Figure 5.12
An instance of a shadow orphan. It is difficult to associate the shadow with the shrub that casts it. Barbizon, September 2016. Image credit: RC.

5.2.4 Common Fate

Whenever an object moves, its cast shadow moves with it. This provides a strong cue to ownership. In a scene with several objects, each moving independently, the common motion of each object and its shadow should allow the proper solution of ownership, even if other factors, like proximity, suggest different solutions. This contribution of common fate has not yet been addressed in vision research on disambiguating the ownership of multiple shadows.

Objects own the shadows that move with them.

5.3 Consequences of the Rules for Ownership

We have described a few of the rules that govern the assignment of ownership. Here we will see that these rules are applied flexibly so as to make the most of the visual scene, sometimes leading to unintended consequences. Specifically, sometimes one rule will predominate over others, and the outcome may lead to erroneous assignments.

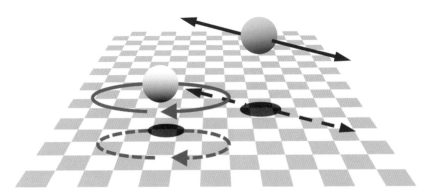

Figure 5.13
Common fate. Two spheres are moving on different trajectories, and their shadows below follow in synchrony. Although proximity may often try to assign the green sphere's shadow to the red sphere, common fate will keep the shadows properly assigned to their respective objects. Image credit: PC.

5.3.1 Proximity and Shadow Capture

Figure 5.3 showed the power of the proximity rule in the limiting case: the object was so close to its shadow that they overlapped (partial hiding). Objects do often partially overlap their shadows, and it is reasonable that the default interpretation is that each object owns the shadow that visually touches it. In these cases, proximity and shape similarity act together to assign ownership. We see an example of the strong synergy between proximity and similarity in one of the earliest representations of a shadow in early Renaissance art (fig. 5.14).

However, the nearest shadow is not always cast by the object next to it. In these cases, we get to see the relative influence of proximity and shape similarity (fig. 5.15). The conflict is almost always won by proximity. Note that if capture occurs, we may actually infer an incorrect position for the light source (see chap. 6). Figure 5.16 presents an example of erroneous capture, or shadow theft, by proximity.

Many examples of shadow capture come from art history. In the left middle part of *Pharaoh with His Butler and Baker*, Jacopo da Pontormo depicts a shadow neatly cast on a wall (fig. 5.17). Whose shadow is it? An observer's first, unreflective answer might favor the character immediately to the shadow's right, a man ascending the flight of stairs. However, upon inspection, it becomes evident that the shadow's profile matches the profile

Figure 5.14
Proximity and similarity act together in this early and rare example of a full cast shadow. Incidentally, observe how neither the rays projected from below nor the attached shadows of the hands are consistent with the position of the cast shadow. Giovanni di Paolo (c. 1402–1482), *The Entombment*, 1426, and detail. Image credit: Walters Art Museum, Baltimore.

of the statue to the left. Indeed, the man on the stairs has his own shadow partially visible to his right. *Shadow capture* provides an incorrect solution to the shadow ownership problem.

Of course, in these conflicts of proximity and similarity, a solution supported by both may still be wrong (fig. 5.18).

In a cluttered image like that of the bottles in figure 5.2, assigning ownership may be intractable. In some cases, proximity and connectedness may contrive to create illusory configurations of objects and shadows. In figure 5.19, what is a shadow of what?

In the scene, some shadows falling across the separate wood joists get grouped together because of their collinearity and may seem to be a single shadow of one overlying wood beam. However, they are not projected by

Figure 5.15
Does the shadow belong to the object that is closer to it or covers it, or does it belong to the object that is more similar to it? Image credit: RC.

Figure 5.16
In a cluttered image, it is difficult to decide who owns which shadow, so the visual system takes a shortcut. The leftmost character captures the shadow of the middle character. Image credit: RC.

Figure 5.17
Jacopo da Pontormo, 1494–1556/7, *Pharaoh with His Butler and Baker*, ca. 1515.
London, National Gallery. Bought with the aid of the National Art Collections Fund
(Eugene Cremetti Fund), 1979.

the same object: an illusion of ownership is produced. The green line on
the right indicates the position of one of the partly hidden beams and the
complex solution to the shadow ownership problem for that beam, where
each individual cast shadow is parallel and almost vertical but not collinear. Instead, illusory beams are tacitly reconstructed by aligned shadows
belonging to different actual beams (e.g., red line on the right in fig. 5.19).

These examples of shadow capture have been based on a conflict of
proximity with similarity. Properties of connectedness and common fate
are also capable of overruling proximity to assign ownership to shadows
that are not the closest. In figure 5.20, the connection between the spheres
and their shadows overrules the proximity solution for two of the spheres

Figure 5.18
Proximity *and* similarity cooperate in delivering an erroneous solution to the shadow ownership problem. The shadow actually belongs to the left hand. Image credit: RC, Paris, October 2016.

Figure 5.19
What is a shadow of what? The green line indicates the position of an actual beam. The red line refers to an illusory beam reconstructed from an accidental alignment of cast shadows of different beams. Image credit: RC, Southern California, May 2014.

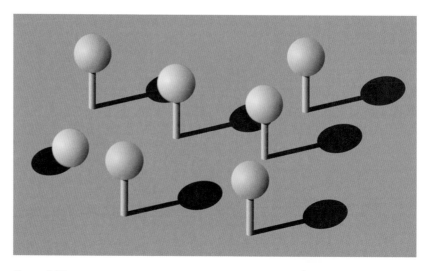

Figure 5.20
Proximity can lose to connectedness in determining shadow ownership. These spheres on stalks cast shadows to the right, and some of the spheres overlap the shadows that are not their own. Despite this proximity cue, the shadows remain owned by the spheres to which they connect and do not appear to lie on the surface (as the sphere on the left does). Part of this ownership resolution may be due to the connectedness and part due to the 3-D elevation of the spheres indicated by the stalks. In either case, these factors overrule proximity in shadow assignment. Image credit: PC.

that overlap shadows that are not their own. Additional support here may come from the similarity in the *direction* of the shadows (as opposed to shape), from the common object-to-shadow separation across several "lollipops," and from the three-dimensional elevation indicated by the stalk.

Finally, when objects are in motion (as described in sec. 5.2.4), another Gestalt grouping principle contributes—common fate—so that in this case we associate shadows with objects that move in step with them, while ignoring possible matches to closer objects. No matter how amorphous a shadow is, and no matter how little it preserves of the visual profile of an object, if the object moves, then its shadow is likely to move with it in the image.

We have shown some special cases where other cues may take precedence over proximity, but in general:

Proximity most often overrules similarity and other cues when they are in conflict, leading to shadow capture.

5.3.2 Shape Similarity: Copycat Shadows

If a bit of similarity between the object and the shadow it casts is good, is a lot of similarity even better? One way to maximize the similarity of the shadow and object shapes is to make the shadow a recognizable copy of the object that casts it—a copycat shadow. This often involves an impossible shadow shape but is nevertheless quite successful, an outcome that indicates the similarity rule is based on crude versions of the object shape and not on actual projections the shadow would have.

Why is a copycat shadow so often impossible? A shadow is a silhouette of the object that cast it, but it is the silhouette seen from the direction of the light, not from the direction of the viewer. So, typically, the shadow

Figure 5.21
Copycat matching is easily satisfied when the object casting the shadow is flat. Image credit: PC, Hanover, New Hampshire, May 2017.

shape and the object's silhouette (as seen by the viewer) do not match. Nevertheless, under particular lighting conditions, the object and shadow profiles can occasionally be quite similar. For instance, when the sun is close to the horizon and at our backs, shadows of people cast on vertical walls can create easily recognizable silhouettes, as opposed to shadows cast on the ground by overhead sunlight, which collapse the shadow into a silhouette of an unfamiliar top view. Thus a shadow that imitates the *visible* profile of the object—a *copycat shadow*—should maximize the contribution of shape similarity to the ownership problem if the analysis of similarity is based on simple two-dimensional profiles. Whereas shadow capture relies mostly on proximity, the copycat solution relies on similarity and usually works quite well, so well that artists will often erroneously exaggerate the similarity between a shadow and the visible profile of the object that casts it.

In this capacity, the copycat solution works in the manner of caricatures that exaggerate the distinctive features of a face well beyond an accurate representation but are nevertheless often very easy to recognize. The peak shift phenomenon seen in herring gull chicks illustrates the phenomenon of perceptual caricature. The chicks peck at the red spot on their mother's beak to induce her to regurgitate their next meal. Niko Tinbergen showed that the chicks responded to a yellow-colored stick with a red strip painted on its side. Moreover, if presented with three stripes instead of one, the chicks pecked even more frequently. Clearly the sticks did not accurately represent the mother's beak, but the trigger feature was the red spot, and the more of it, the better. The same seems to hold true for the shape similarity cue to shadow ownership.

We find one of the earliest examples of the copycat method in an illumination representing a biblical episode. In the Old Testament book of Joshua (2:15), the Gentile prostitute Rahab helps two men sent by Joshua as spies to Jericho to escape on a rope from a window of her house. The episode was illustrated in the so-called *Visconti Hours* by Belbello da Pavia, in the mid-fifteenth century (fig. 5.22).

The artist has depicted the shadows of the two spies on the left-hand wall, although the shadow of the left-hand spy is partially hidden. These shadows copy the visible silhouettes of the pair of spies. But, of course, the light that would cast that profile would be from the viewer's direction, producing a shadow behind the spies, not to the side. The shadow is therefore physically incorrect, but our eye hardly notices the mistake. However,

Figure 5.22
Redrawn version of Belbello da Pavia, *The Spies of Jericho Escape*, in the *Visconti Hours*, ca. 1420–1425. Florence, Biblioteca Nazionale, BR 397, LF 22. Note the copycat shadow of the right-hand spy on the left wall. Image credit: PC and RC.

although the artist has used a copycat method, which should link the right-hand shadow to the right-hand spy by similarity, proximity still rules, and the right-hand shadow appears to belong to the left-hand spy despite the mismatched head, hat, and beard shapes.

Figure 5.23 shows another example of the copycat influence on ownership. On the left, the shadow arrangement matches our view of the similar object shapes, and shadow ownership matches each object to its copycat shadow, even though some of these matches ignore proximity. However, this shadow arrangement is physically incorrect: to cast a shadow that

Figure 5.23
Copycat (*left*) and real shadows (*right*) of four solid objects. In the copycat case, the four objects appear to be at the same distance from the light gray wall. In the real case, the objects appear to be at different distances from the wall. Image credit: Jelena Arsenic and Svetislava Isakov, 2005.

matches the three-dimensional object profile, the light source should be close to the viewpoint, whereas in figure 5.23, the light is clearly coming from the left side. The physically correct shadows (fig. 5.23, right) are instead organized vertically, and here the ownership of the shadows is harder to work out. Note that this shows that it is not the one-to-one match of object and shadow shapes that overcomes proximity, because the shadows in the right-hand panel are also close copies of the object's shapes. Instead the two-dimensional layout of the shadows plays a strong role as well, suggesting that this similarity mechanism is insensitive to three-dimensional structure.

A dramatic example of copycat shadow shows up in a twentieth-century masterpiece, Magritte's *Empire of Light* (of which several versions exist). The shadow cast by the lamp casing is not geometrically correct, yet it is not only acceptable but preferable to the correct shadow because it matches the casing's shape. The painting is usually taken as an illustration of a "paradoxical" treatment of light: it is as if a daylight sky was juxtaposed with a dark night scene. The painting's lower half is lit by a feeble streetlamp; the shadow of the streetlamp's casing is cast on the wall of the building (see the two dark oblique lines visible next to the central windows). Now, lamps such as the one Magritte painted are typically assembled from glass panes

 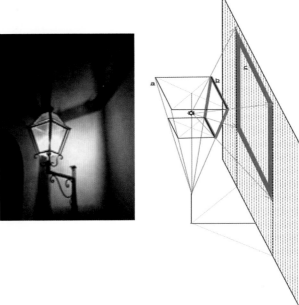

Figure 5.24

Left: Tapered casing shadows in Magritte's *L'empire des lumières*. *Middle*: A real-life example of a streetlamp casting shadows of its case on a flat wall. Note that the shadow has something implausible about it and appears to fail the ownership test. *Right*: The diagram shows that the casing's shadows should be vertical, not tapered. Image credits: Left: René Magritte (1898–1967), *L'empire des lumières*. The Solomon R. Guggenheim Foundation Peggy Guggenheim Collection, Venice, 1976. 2017. Herscovici, London/ Artists Rights Society (ARS), New York. Middle: RC, Maiano, Italy, 2005. Right: RC.

of equal size and shape. As these are isosceles trapezoids, the lamp is just a truncated pyramid, and the shadows of the case edges—projected from the center of the lamp—should appear as parallel vertical lines. Indeed, in real-life situations, similar streetlamps cast "straight" shadows that are visually awkward (fig. 5.24, right). However, in Magritte's painting, the shadow, as it stands, is both recognized as a shadow and assigned to the appropriate object—the lamp—because the shadow's tapered geometry matches that of the lamp casing.

In some cases, we are presented with all three actors (light source, object, and surface) to help us assess the plausibility of the representation. In Vasari's *The Origins of Painting*, the light source is in a position such that it could not possibly cast the shadow of the painter's profile on the wall (fig. 5.25). Actually, no light source is positioned anywhere in the scene that

Figure 5.25
Did painting originate from copying a shadow? Giorgio Vasari (1511–1574), *The Origins of Painting*, 1573. Florence, Casa Vasari. Image credit: Photograph by Sailko.

can possibly cast that very shadow, as the painter's profile exactly matches the shadow's outline. (Sure enough, the real shadow from that position would have been a poor portrait of the painter.) Observe that the painter's face is in the shade, which means that his profile is definitely not casting his shadow as depicted. Nevertheless, the shadow seems appropriate at first glance.

Copycat shadows demonstrate the power of similarity in influencing shadow ownership.

5.3.3 Shadow Abandonment

Static, free-floating shadows (fig. 5.26) are perceptually puzzling. Flying objects are normally not static; they need movement to fly. Static objects, on the other hand, are normally attached to the ground, so their shadows cannot be free-floating. They must be connected to other shadows or at least to the edge of the visible scene, as would be the case for a shadow cast into a scene by a tree or a traffic sign that lay outside the field of view. (Clouds may be an exception, given their slow speed, but their shadows are typically so large that they do not trigger an ownership analysis—unless you see them from an airplane.) These regularities are probably engraved in the mental encyclopedia for shadows, so that our surprise or discomfort in seeing a static, free-floating shadow may even lead us to reject it as a shadow.

Unsolved ownership produces shadow abandonment with the possible loss of shadow labeling.

5.4 Conclusions

The shadow ownership problem is solved by a set of powerful heuristics. As a default case, proximity (with contact as a limiting case) is the computational glue connecting an object to a nearby shadow, often ignoring shape inconsistencies. This heuristic assumes that we somehow first evaluate dark regions as potential shadows (as discussed in chap. 4) before assigning ownership. More generally, the Gestalt factors of proximity, shape, connection, and common motion are responsible for deciding who is a shadow of whom. These separate factors may agree or disagree in any given image, and their use can generate illusions of ownership like shadow capture. The

Figure 5.26
The thin cords holding the banner cast invisible shadows. The shadow of the banner appears to be free-floating. Because it is static, it may lose its shadow character. Image credit: RC, Strasbourg, 2005.

process of determining ownership is undoubtedly more interactive and flexible than the simple linear sequence we have described here. Some areas may be labeled as shadows only once an appropriate owner has been found, and others only if the inferences of shadow, object location, and shape support a good or likely interpretation of the scene.

To sum up, shadow ownership is strongly influenced by Gestalt principles, the same ones that oversee the construction of objects: proximity, connectedness, shape similarity, common fate, and so on.

Proximity of a dark region is a strong cue to shadow ownership.
Ownership of anchoring shadows requires a good connection.
Shadow ownership requires that the shadow shape be approximately conformal.
Objects own the shadows that move with them.

However, these principles are applied in a simplified manner, and conflicts between them can lead to misattribution of ownership, generating the phenomenon of shadow capture or theft.

Proximity most often overrules similarity when they are in conflict, leading to shadow capture. However, other factors like connectedness and common fate can overcome proximity, leading to correct assignment of ownership.

Copycat shadows demonstrate the power of similarity in influencing shadow ownership. It is rare that a shadow's shape matches the visible profile of the object that casts it (only when the object is itself flat), and yet when this happens, the assumption of ownership is very strong.

Unsolved ownership produces shadow abandonment. In some cases, this leads to a loss of shadow labeling.

Mission accomplished? Not yet: as the example of abandoned shadows shows, not everything that is a shadow is seen as a shadow. The next chapter examines how the visual system chooses what is or is not a shadow.

Further Research Questions

Shadow ownership creates visual "molecules" composed of an object and its shadow.

1. Is ownership attribution a special case of Gestalt composition specific to shadows? Or do both tap into the same computational resources?
2. Shadow ownership combines objects and their shadows; it is transcategorical. In the standard Gestalt composition of images, each atom within a grouping has the same status as the others. Does this difference affect the speed or reliability of shadow–object ownership links?
3. What other cases can be documented of transcategorical molecules? We describe later on (secs. 8.2 and 9.2.2.2) light ownership and reflection ownership, but are there other cases? Echoes? Overtones? Is there an event ownership problem?

Larry Kagan, *Mosquito*, 2007. Steel and shadow. The mosquito image is a shadow of the wires, from a light above. Wire sections are assembled and welded to generate the surprising shadow, matching a sketch draft on the wall; the sketch is then erased, and only the mosquito-like shadow is left. In this case, the shadows of the wires also create what appears to be a shadow of the illusory mosquito shadow. Image credit: Larry Kagan.

6 Labeling Shadows

6.1 Shadow Information Only Available Once Shadows Are Labeled

The solution to the ownership problem enables shadows to accomplish their mission: knowing what is a shadow of which object is a prerequisite to using that shadow as an information vector for that object. But before a human observer or a computer vision system can extract any scene information carried by shadows, candidate regions must be evaluated to determine which can be accepted as shadows. Borrowing a term from computer vision that is also common in vision research, we call this step in the visual system "labeling." Only once a region is labeled as a shadow can it contribute to the recovery of scene structure. It is then "demoted"; that is, the shadow contours are subsequently ignored in segmenting object and scene regions. However, although filtered out for object processing, the labeled shadows are nevertheless perceived as shadows: they have subjective shadow *character*.

The range of information that shadows carry indicates that many optical properties could be used to label shadows. However, probably for purposes of speed, biological visual systems have chosen to use only a small subset of these properties. The goal of this chapter is to describe how the visual system accomplishes this labeling: what are the properties used by the visual brain to decide that a region in a scene is a shadow. The matter is convoluted, and we are about to embark on a bumpy ride. In particular, we will need to go back and forth between what we know of the shadow mission and ownership and what we are about to know about shadow labeling. This back-and-forth reflects the complexity of the computational architecture of shadow vision.

Figure 6.1

Painted dark regions below the car and the man pass the test for human vision and create an illusion that both are floating above the pavement. In the video, the car then drives away to reveal the deception. The "shadow" shapes and location do not correspond to the illumination; notice there is no direct light falling on the car or the man, and no corresponding highlights on the car. These physical properties are not part of the set of rules used by the human visual system. Image credit: Honda.

6.1.1   Some Shadows Contribute to the Understanding of a Scene but Do Not Need to Be Labeled as Shadows

Before proceeding, we should ask whether the contribution of shadows to scene understanding always requires labeling the shadows as shadows. In surface perception, for example, shadow contours may provide information about the shape of surfaces on which they fall. In figure 6.2, the shadows act as permanent surface marks, and their parallel texture helps define the three-dimensional articulation of the surface on which they are cast. We can tell that they act as surface marks, and not in their capacity as shadows, because *we can reverse their contrast and still recover the surface shape from these "shadow" borders.* That these contours arise from shadows is not relevant for their use in retrieving three-dimensionality. In this sense, they act in the same way as a zebra's stripes (fig. 6.2, bottom) to reveal three-dimensional shape (although in the case of cast shadow contours, they also change as

Figure 6.2

Top left: Lines created by shadows correlate with the shape of the object on which they are cast. Inverting the contrast (*top right*) does not eliminate the shadows' contribution to three-dimensional recovery. *Bottom*: The pigmentation lines in these black-and-white-striped zebras contribute to three-dimensional recovery in the same way. Image credits: Top: RC, Versilia, July 2014. Bottom: John Storr.

the object moves through the shadow field, giving more 3-D information as they do). We should point out that these parallel shadows are infrequent or absent in natural environments, where the typically dappled, random shadow shapes do little to reveal the shape of the object they fall on. This dappled pattern is believed to be widely used as camouflage by many animals, although there are controversies about its effectiveness.

6.1.2   The General Case: When Shadow Labeling Makes a Difference

In general, however, it is important that the shadow be processed as a shadow to assist in the reconstruction of the three-dimensional structure of a scene. A nice example of the difference between perceiving a dark area in the image as dark pigment, and perceiving it as a shadow, and of how each impression can create different spatial organizations, can be found in Salvador Dalí's *Slave Market with the Disappearing Bust of Voltaire* (fig. 6.3).

Dalí's painting—a complex statement about the presence/absence of Voltaire, a declared antislavery thinker, in the scene of a slave market, and

Figure 6.3
Salvador Dalí (1904–1989), *Slave Market with the Disappearing Bust of Voltaire*, 1940. Oil on Canvas. © Salvador Dalí Fundació Gala–Salvador Dalí (VEGAP) 2015. Collection of the Salvador Dali Museum, Inc., St. Petersburg, Florida, 2015. © Salvador Dalí Museum, Inc.

about the way in which Voltaire's philosophy enslaved the mind—exploits the ambiguity of black areas in the image, areas that could just as well be reflectance features or illumination features. When you look at the painting, you can see either a bust of the writer or two merchants under an arch in the wall. If you see the bust, then you interpret the dark spots as shadows on a white stone sculpture. The zones under Voltaire's eyebrows (fig. 6.4), being in the shade because of a strong light from above, are seen as concavities. If, on the other hand, you notice the slave merchants instead, dressed in traditional aprons and floppy hats, then those same zones correspond to the permanent black fabric of their hats and suggest no concavity. The two competing interpretations affect the spatial layout of the scene,

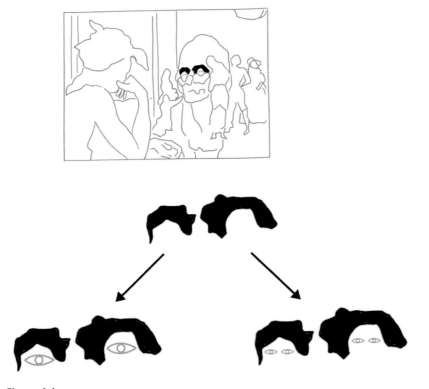

Figure 6.4
Black spots on the surface of the painting are interpreted either as shaded concavities of the white statue or as nonshaded black pieces of fabric, thereby contributing in different ways to the three-dimensional interpretation of the image. Image credit: RC.

and quite different possibilities are opened by labeling the relevant areas as shadows as opposed to not doing so. More generally, if you see something as a shadow, you should expect it to make a *specific* contribution to the reconstruction of the three-dimensional structure of the scene.

Note, however, that labeling the shadow and extracting the available information do not constitute a simple sequence. In many cases, as in figure 6.3, the regions that may be shadows have alternative interpretations, leading to alternative object and scene structure. Typically, one scene structure is more plausible than the alternatives, but in figure 6.3 the two choices are equally plausible and can switch back and forth. The point, of course, is that some properties of shadows (see sec. 6.3) can veto an area as a shadow, but many dark regions that are not shadows will pass the tests. It is then a more complex inference that selects whether an area is a shadow based on the plausibility of the scene structure it supports.

6.1.3   Shadow Label and Shadow Character

Shadows' perceptual functions can only be accomplished once they are labeled by the visual system. This first stage of labeling is a rapid, unconscious process of impressive sophistication and power that is the subject of this chapter. Once it performs the labeling, the visual system can access and use the information about scene structure and illumination that shadows carry, and then "demote" the shadows (chap. 4). However, our appreciating that certain areas *appear* to be shadows—have *shadow character*—is an optional, conscious observation that is incidental to the effects that the labeled shadows have already had on our perception.

If a critical property of a shadow is violated, it will lose both its labeling and character as a shadow—and in the following sections, we take advantage of this loss to determine these critical properties.

Shadow labeling is distinct from shadow character: it is possible to have a region that acts like a shadow but does not look like one. For instance, in some conditions, a dark region will trigger compensation for illumination: the decrease in perceived illumination indicates that it has been dealt with as a shadow, even if it does not look subjectively like a shadow. In figure 6.5, for example, shadow character is absent or very weak, but an illumination variation (or possibly a superimposed dark filter) is discounted nevertheless. This phenomenon mimics the effects of a real shadow, as the parts of the cubes along the dark horizontal strip seem to receive less illumination than

A cautionary note: Language introduces biases in what we say when we talk about anything, and about shadows in particular. This affects subjective reporting. We should accept that some concepts are not lexicalized: there are mental categories or descriptors that do not find their way into verbal expression. Furthermore, different languages may lexicalize the same range of phenomena differently. (For instance, Italian does not have a lexical representation for the distinction between "shadow" and "shade.")

Figure 6.5
Illumination variation is a more general effect than the presence of a shadow. All the diamonds have the same luminance, as demonstrated by the gray borders, but the diamonds in the dark band look much brighter than those in the light band. The X-junctions between the horizontal borders and the vertical cube edges support an assumption of an illumination difference (or transparency) that leads to a correction in apparent lightness. Image credit: PC, based on Logvinenko (1999).

the parts along the lighter strip. The rows of diamonds—the top faces of the cubes—appear dark in the light strip and light in the dark strip, but they have the same luminance in the image (check the matches to the vertical borders on the left and right). Thus certain rules appear to trigger shadow labeling, that is, treating certain superficial features as accidents of illumination, which in turn induces an illumination compensation, without necessarily generating the impression that the labeled regions are shadows.

Shadow labeling and character can be applied both to real shadows and to nonshadows—as they can also be applied to representations of shadows in photographs and paintings (mere assemblages of patches of ink or paint) and movies (patches of different illumination). Equally important, though, are the conditions that break shadow labeling and character for legitimate shadows. We call this *shadow disruption* and describe and use it in the next section. The simplest way to understand the conditions that make us perceive a shadow as a shadow is to find the contexts in which an actual shadow is *not* perceived as a shadow, that is, conditions of shadow disruption.

We do not know the exact conditions that lead to the disruption or survival of shadow labeling. We begin by presenting a list of properties that are always true of real shadows, before proceeding to test them one by one, to sort out which can, and which cannot, be ignored in the acceptance of a region as a shadow.

6.1.4 Physical Properties That Define a Shadow

Here is a short summary of the physical properties that are always true of real shadows—no matter if we perceive them as shadows or not. The key point is that as the shadow falls across a surface, we do not expect changes in any features across the border except, of course, the illumination itself (fig. 6.6). The following are four types of properties that are true of shadows, expressed in the language of laws and constraints that we established in chapters 1 and 2.

1. *Discontinuity in light and color between the shadow and nonshadow region.*

    1a. Illumination decreases in the shadow region: *shadows are darker.*

    1b. *Some* color changes can occur across the shadow border if there are two light sources. The color in the surround includes the color of any ambient light that falls in the shadow. For example, an outdoor shadow may be bluish from the sky light, but the surround illumination has to include the sky light as well as the direct yellow sunlight, producing white.

2. *Continuity of surface properties from the shadow to the nonshadow region.* No changes in surface properties are in the norm aligned with the shadow border.

    2a. Contours in the surface texture continue across the shadow border uninterrupted, forming characteristic X-junctions with the shadow border. The shadow does not block the surface contours across which it falls.

    2b. *The nature and the contrast of the surface texture, if any, are the same on both sides of the shadow border.*

This is a key structural feature of shadows. Figure 6.6 and the image in the margin explain the property of X-junctions.

3. *Two-dimensionality.* The shadow is only an illumination change on the receiving surface. It *does not have any volume of its own.*

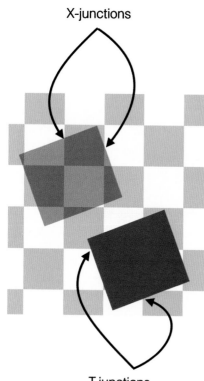

X-junctions

T-junctions

X-junctions suggest transparency or shadowness; T-junctions indicate occlusion. Image credit: PC.

Figure 6.6
The birth of a shadow. Same color, same type of flat shape, but only one of the gray areas, the one whose contours make X-junctions with all other contours, is a shadow. Courtyard of California Academy of Sciences, San Francisco, May 2014. Image credit: RC.

4. *Shape and position.* The shadow shape is determined by the shape of the object casting the shadow, by the distance and direction of the light source, by the distance between the object and the receiving surface, and by the topology (orientation, curvature) of the receiving surface. All the shadows from each light source fall consistently along the illumination direction away from the source. When the object contacts the surface on which its shadow falls, the shadow border meets the object border where it is tangent to the illumination direction.

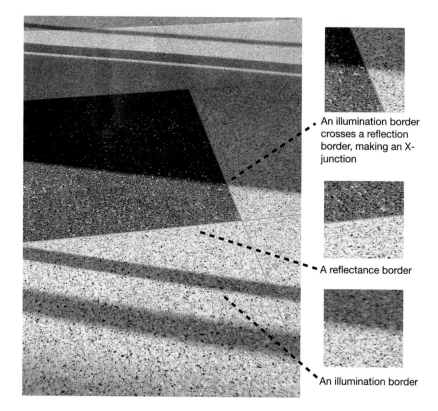

An illumination border crosses a reflection border, making an X-junction

A reflectance border

An illumination border

Figure 6.7
The cast shadows fall over the tiles of different colors and should change the illumination and to some extent the luminance (because of the bluish sky light falling into the shadow) at the border. But nothing else should change at the border. Note the X-junctions between the shadow border and the borders of the tile. Image credit: RC.

Although complex, these interactions are straightforward and are used to render scenes realistically by computer-based rendering programs.

As we will see—and this is the key point of this chapter—*the human visual system checks only some of these properties to label shadows*: our visual system does not care about a complete, perfect representation of the shadow world. Specifically, for the discontinuities, only the brightness property holds: shadows must be darker but can have arbitrary colors. Of the continuity properties, alignment of changes other than illumination or color at the border, and failure of surface contours to cross the shadow border, can veto a shadow labeling in some circumstances, but not all. The two-dimensional

Figure 6.8
Contrast reversal destroys shadow character. The negative image of a shadow looks like some white keys or white paint. This shows that a shadow must be darker than its surround. Image credit: RC, Barbizon, August 2016.

property is critical: a shadow region should not have object-like features of its own, such as volume. Finally, only a few of the shape properties seem to be used: a region can be taken as a shadow even if its shape does not match the required shape and even (although not always) if its direction is inconsistent with that of other shadows in the image. The only shape property that is invariably critical is that when an object is sitting on a surface, its cast shadow must meet the object's contours appropriately.

6.1.5 Testing the Properties of Shadows

How can we determine which of these properties are used by human observers? We can just ask whether a region that violates one of the properties no longer looks like a shadow, and we will often take this simplest subjective *shadow disruption* approach: we notice that when shadow character is lost, the area no longer *looks* like a shadow (fig. 6.8).

A subjective judgment about shadow character may be difficult or ambiguous in some cases. A more objective test is to ask whether a region *acts* like a shadow (chap. 4)—for example, does a shadow generate an impression of depth (fig. 6.9). We are still asking for a subjective judgment, but the

Figure 6.9

Left: When the patch occluded by the blue square is darker than its surround, we take it as a shadow, creating an impression of a separation in depth between the square and the background. *Right*: However, if the same patch is lighter than its surround, we get no impression of a shadow and little or none of depth. Image credit: PC.

effects of biases and aesthetic influences on a judgment of shadow character are no longer an issue. Does the blue square appear to float in front of the background? Yes. Does it also appear to float in front on the right? No. Conclusion: shadows need to be darker than their surrounds. We call this an "objective" test, realizing that it is still grounded on a subjective judgment, of the effect of the shadow labeling rather than of the shadow character itself.

Here we capitalize on what we know of the shadow mission that was the focus of chapter 4. We rehearse some of the points of that chapter here to explain the objective tests we will use. The logic is to take one of the scene features that shadows support, such as surface lightness (fig. 6.8), depth (fig. 6.9), or demotion (fig. 6.10; shadows are not very salient, and it is hard to find an odd one). With each of these techniques, we can ask whether violating a shadow property (like being darker) eliminates the lightness, depth, or demotion that shadows should support.

6.2 Mandatory Shadow Properties

If a region in a scene supports one of the shadow-triggered scene features we have mentioned (lightness, layout, demotion), then that region is likely

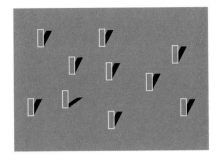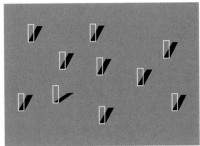

Figure 6.10

Left: The odd dark shadow is hard to find, as it is labeled as shadow and thus demoted. *Right*: The dark areas do not conform to the possible shapes that the shadows of the outline bars should have. Since the dark areas are not treated as shadows, the odd one is easier to find. Image credit: PC, adapted from Rensink & Cavanagh, 2004.

to have been treated as a shadow. If violating a shadow property for that region destroys the layout or lightness inference, then this shadow disruption reveals a property that is *mandatory* for shadow labeling. In contrast, if violating the property does not disrupt the shadow labeling, then we have identified a property that could be used by a computer vision system, for example, but is *ignored* by our human visual systems.

Using both these objective shadow disruption tests and simple subjective tests ("does it still look like a shadow?"), we find that only a few of the physical properties of shadows listed in section 6.1.4 turn out to be mandatory; the relevant areas can be the wrong color or can be unrelated to the casting object's shape and still be labeled as shadows (and so support recovery of shadow-dependent information) and have shadow character (look like a shadow). The properties that are necessary are that the shadow is darker, the shadow does not have a contour around it, the shadow contours make X-junctions with background contours, and the shadow has no volume of its own.

Note that having these properties is necessary but not sufficient. A region may pass all these tests but end up classified as something else—not a shadow but a dark pigment, stain, or transparency—because that classification leads to a more plausible scene interpretation. So these properties can veto a shadow interpretation if broken, but constitute only a first step to shadowness if they hold.

Why do only some properties apply, and why are some true properties of shadows not considered at all? Once more, speed and economy rule: in

Figure 6.11
Shadows are regions where the illumination is blocked, so necessarily they are darker than adjacent regions where the illumination is not blocked. However, this rule applies only locally along the shadow contour, as we will see in the following section. Image credit: PC, Marrakesh, December 2014.

the pursuit of the speed required to survive in a rapidly changing environment, our visual system found a subset of properties to be sufficient in most cases; these emerged as the working shadow-labeling toolbox for biological vision. This reduced set of properties supports much of the richness of our visual culture that would not be possible if our visual systems had to check all the physical properties like those used in computer vision and computer graphics: some of them are so computationally demanding that our visual systems would never arrive at appropriate shadow labeling within a useful time interval.

Let us start with the most obvious property of all: shadows must be darker.

6.2.1 A Shadow Must Be Darker—but Only at Its Edge

Clearly the most reliable property of real shadows is that they are darker than their surrounds. This is physically their definition, and we would be quite surprised if shadows lighter than their surrounds would still act as shadows rather than, say, reflections that are always physically lighter than their surround. Given the number of other shadow properties that are not necessary for regions to be perceived as shadows, it is reassuring that, at least for human observers, inferences based on shadows require that they be darker—with a few caveats. Note that the darkness property is not specific to shadows. For instance, water on a surface filters out light and creates a dark area that may, or may not, be interpreted as a shadow (fig. 6.12).

Both shadow character and shadow labeling are strongly associated with the darkness property. Images of shadows that are contrast reversed do not look like images of shadows anymore; in figure 6.8, for example, the shadow of the keys looked like light paint when the contrast was reversed. An objective test is to reverse the contrast of shadows and show that this destroys the shadow labeling and the depth inferences that are based on it. The example of figure 6.13 shows a positive and negative contrast version of a two-tone image of a face (see the version in the margin if the

When a shadow forms contours that reveal the object's structure in the way that surface markings would, contrast reversal does not veto recovery of depth.

Figure 6.12
Shadows, wet surfaces, and black stains are in the norm darker than their surrounds. Adding object images—even cartoonish ones—may make us perceive the wet area as a shadow. Image credit: RC, Paris, May 2017.

Figure 6.13
Contrast reversal destroys a shadow's contribution to the 3-D interpretation of an image. *Left*: The image can be seen as a man's face illuminated from the left. *Right*: The identical image, in reverse contrast, does not support the same sense of depth, though we know what it is and where the eyes and nose are placed. Image credit: Cavanagh and Kennedy (2000).

To assist recognition: two-tone face with eyes, nose, and mouth superimposed.

face is hard to see). As these images have only two levels of contrast (light and dark), some of the dark regions could be dark pigment, but others are shadows under the nose, cheeks, and eyebrows. Once the dark regions are labeled as shadows, the appropriate depth interpretation of the face emerges. However, with the contrast reversed, although the same contour information remains, the *depth interpretation* is almost completely lost. This is an objective test of the importance of shadows being darker.

There is an important caveat to the rule that shadows must be darker. In fact, they may actually be lighter in their interior *as long as they are darker at the edge*. This is a critical observation because it tells us that the processes that verify whether a region is a shadow *check the darkness property only locally along the border* and then *propagate the property of "shadow" into the contained region*.

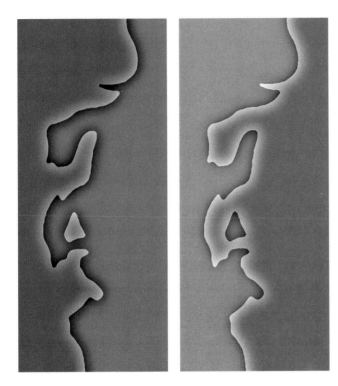

Figure 6.14
The man's face may be more apparent on the left than on the right. *Left*: The brightness inside the original shadow areas on the right of the face is physically lighter than in the original illuminated area on the left. However, because of the appropriate direction of the switch from light to dark at the shadow's edge, it passes the contour test for shadows and supports to some extent the recognition of the face and its depth structure. *Right*: The area inside the shadow regions is physically darker, but the switch at the contour goes the wrong way, from dark to light, making the face and the depth difficult to recover. Image credit: PC.

The face interpretation on the left in figure 6.14 is preserved because all along the border the shadow region is darker than its surround. It makes a lot of sense that the brightness constraint is not checked throughout the entire shadow, since complex textures could randomly have some brighter regions inside the shadow or darker outside it despite the overall illumination differences.

It may seem trivial that shadows must be darker (at least along the contour), but we see a counterexample with shading, the variation of light

Isophotes, lines of equal luminance, over the surface of a statue. Image credit: PC.

reflected from an illuminated surface because of its different orientation with respect to the light source. When an image is reversed in contrast, all the contours remain at the same location; all that has changed is the direction of change in light to dark across the contour. Reversal destroys the shadow-based depth inferences (fig. 6.13, right), but the same does not hold true for shading, where reversal has no effect on the recovery of depth (fig. 6.15). As it happens, the depth information from shading is carried *not by a single boundary*, as it is for a cast shadow, but *by the family of contours of the same luminance* (called isophotes). This family of contours is preserved when the polarity is reversed (as is the shadow contour), and the depth recovered from the shading is preserved as well. We present this to underline the point that depth from shadows could have been recovered independently of the lightness or darkness of the shadow region, but the visual system has chosen not to do so. It makes a different assignment for two-dimensional regions that are lighter than their surround: typically, it treats them as reflections. So, *unlike the case with shading, in the case of cast shadows, a reversal of contrast requires a reversal of labeling.*

Verification at borders. As we have described, the property that shadows must be darker than their surround is too broad. The darkness property must hold along the contour, but one reason that it cannot be applied to

Figure 6.15

The right-hand version of this horse is the reverse-contrast version of the left-hand image. The left-hand image was taken with the light and the camera on the same axis (using a half-silvered mirror), so no cast shadows are visible. The 3-D structure of the horse is unchanged by the contrast reversal, even though no physical lighting could produce the shading on the right-hand image (although it is possible if the surface material is seen not as matte but velvet-like). Thus depth from shading, unlike depth from shadows, is indifferent to the polarity of contrast. Image credit: PC.

all areas of shadow and surround is that shadows often fall on surfaces with arbitrary variations in reflectance. Some light regions within the shadow may have higher luminance than some dark regions outside. The shadow area does not have to be *uniformly* darker than its surround. This underlines the critical importance of the luminance change being tested *specifically at the border*. At the border, each point just outside the shadow must be brighter than each immediately adjacent point just inside the shadow.

Now, shadow borders themselves may cross areas of different reflectance. Thus the luminances within the shadow areas can be extremely different, even close to the border. But wherever the shadow border crosses reflectance edges in the surface texture, it sets up ratios that must respect the constant contrast across the illumination boundary, no matter what reflectance it falls on (although with artificial light, the luminance and so the ratio can change with distance from the source). In figure 6.16 on the left, the ratio of the luminance values of adjacent surfaces A and B must be the same as the ratio of other adjacent points along the contour A′ and B′, *as they are all caused by the same change in lighting*. And that ratio must be less than 1.0 (the shadow side must be darker). If this ratio holds along the border, it is appropriate for the presence of a shadow. If the cast shadow region is made lighter than the unshadowed area (fig. 6.16, middle), the ratio may still be a constant at all points along the border, but it becomes greater than 1, and the shadow character disappears, sometimes creating transparency effects (see chap. 9). Violations of the ratio requirement may also create the effect of a change in pigments—different paints—rather than a change in illumination (fig. 6.16, right).

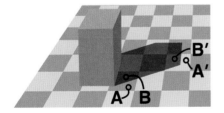 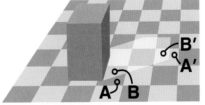 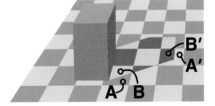

Figure 6.16
Left: The contrast between adjacent points across the shadow border must be equal at all locations: the A:B luminance ratio should be equal to the A′:B′ luminance ratio. *Middle*: The same-ratio rule is respected, but the ratio is greater than 1, hence creating possible overlay of transparent material but no shadow. *Right*: Violations of the ratio rule create the effect of different pigments. Image credit: PC.

In sum, shadows have to be darker, but the verification of this property appears to happen at the border. This is both a highly reasonable constraint, given the unpredictable nature of reflectances in complexly textured surfaces, and a frugal choice, as it limits the necessary computations to extremely local regions at the border. Thus:

Shadows must be darker than their surround along the border.
The ratio of luminances is consistent along the edge.
The shadow label is propagated into the region bounded by the shadow border.

6.2.2 Shadows Must Cross Surface Contours, Not Align with Them

Real shadows fade from dark to the background level smoothly (although sometimes sharply). So shadows have a boundary, and that boundary *generally crosses surface contours, making X-junctions* (fig. 6.6; see also fig. 6.16,

Figure 6.17
Highlighted contours prevent shadows from contributing to three-dimensional recognition. Image credits: Left: Giorgio Kienerk, *Il Sorriso* (detail), ca. 1900. Museo Giorgio Kienerk. Right: PC.

left). The shadow boundary may occasionally be aligned with (and hence not cross) a portion of a surface contour, but a contour aligned with the shadow border over a considerable extent is a sure way to defeat a shadow interpretation—it is too much of a coincidence for the locations of shadow borders. This can be seen as an objective test by registering the loss of three-dimensionality when an outline is added to a shadow area. The artist Giorgio Kienerk demonstrated this effect in his *Sorrisi* (Smiles) series by placing contours along the light-to-dark borders (fig. 6.17, left). When the contour is removed (fig. 6.18, left), the depth becomes more evident. The same effect is obtained with a high-contrast version of a man's face (fig. 6.17, right and fig. 6.18, right).

As another objective test of the disruptive power of an aligned contour, visual search tasks for an odd shadow direction that are difficult when the

Figure 6.18
Without the extra contours, the appropriate portions of the dark areas are interpreted as shadows and support the recovery of the three-dimensional face structure. Image credits: Left: PC, adapted from Kienerk, *Il Sorriso*. Right: Cavanagh and Kennedy (2000).

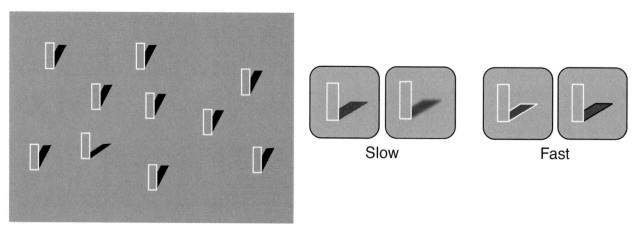

Figure 6.19

Finding an odd shadow among others, as on the left, is slower when the region adjacent to the vertical bar is dark with sharp or blurry edge. However, searching for the odd item is fast if a contour surrounds the region that would otherwise be taken as a shadow. Image credit: PC, adapted from Rensink and Cavanagh (2004).

New York City artist Michael Neff draws chalk outlines around shadows. Image credit: Michael Neff.

shadow is darker than the background become easier when that shadow has an outline (fig. 6.19). The regions are then no longer demoted as shadows, and their properties are easier to access.

The subjective shadow character is also powerfully affected by an outline. Take a small object, put it in direct light, and have it cast a shadow on a white sheet of paper. Now draw a thick black line exactly matching the profile of the shadow. The area where the shadow is located does not look any more like a white area in the shade; it appears to have a color of its own. The color of the area appears solid, as if it were overlaid on the sheet (fig. 6.20). We call this the Leonardo effect, as Leonardo da Vinci appears to have been the first to describe it:

> When you represent in your work shadows which you can only discern with difficulty, and of which you cannot distinguish the edges, so that you apprehend them confusedly, you must not make them sharp or definite lest your work should have a wooden effect.

At the end of the nineteenth-century physiologist Ewald Hering offered an explanation of this phenomenon: the added outline of a shadow (a "Hering line," as we may call it) destroys the penumbra, which he took as the phenomenological mark of the shadow character. The idea is ecologically valid, as most shadows are accompanied by a penumbral area, but is it the end of

Figure 6.20
The Leonardo effect: Outlining a shadow destroys its subjective character. Instead of a monochromatic sheet of paper, partly shaded (*left*), we experience (*right*) three colors: bluish gray, black (outline), and white. Image credit: RC.

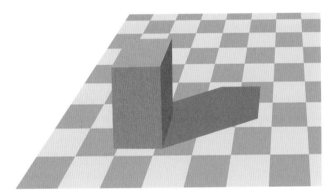

Figure 6.21
An opaque shadow acquires a life of its own and becomes a carpet that is overlaid on a surface. The T-junctions at its border degrade the shadow interpretation. Image credit: PC.

the story? A few decades later, Carl Bühler disagreed. He pointed out that if you draw lines or textures *inside* the shadow while leaving the penumbra intact, the shadowy character disappears as well. This provides evidence that the region is in fact a separate opaque surface or a surface marking. Thus the penumbra does not seem to be a critical property.

Indeed, any sort of feature that ends at the shadow border can veto or diminish the shadow. For example, if a candidate shadow region blocks reflectance borders, creating T-junctions (fig. 6.21), it will appear opaque

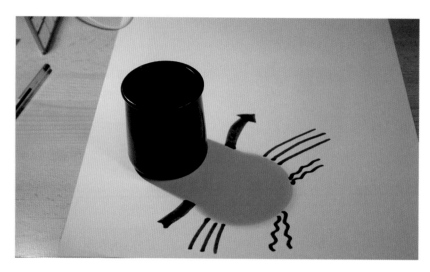

Figure 6.22
Bühler's lines turn the shadow into an opaque patch. Interrupted lines matching the boundary of the shadow indicate that it is not an accidental illumination boundary. Image credit: RC, Paris, March 2017.

and lose its shadow label and character. In figure 6.22, the shadow appears to interrupt several lines, creating the impression that the shaded region is an opaque surface with lines continuing amodally behind it. Examples in art are rare, as they are easily seen to degrade the shadows in the image (see fig. 6.23). In all these cases, the unlikeliness of the coincidence of a shadow with a stable reflectance feature of the environment vetoes the shadow interpretation. An exception occurs when the direct illumination is the sole light source and no light falls in the cast shadow. In this case, no details are visible in the shadow, and shadow character is preserved.

An objective test for opaque regions as candidate shadows is once more the visual search task where acceptable shadows are demoted and their orientation is harder to judge. Here the regions with T-junctions around their border are easier to find (fig. 6.24, rightmost two examples), even though they are suitably darker than their surround, indicating that they are not taken as a shadow. We can summarize all these findings in the following critical property:

The shadow border must cross surface contours, not coincide with them.

Figure 6.23

Opaque shadows block the reflectance differences they may fall on and do not look like shadows. In the left image, note how the "shadows" on the bridge mask the bridge features, while in the right image the "shadow" even covers another shadow. Image credits: Left: Hunting mosaics, Villa Romana del Casale, Piazza Armerina, third–fourth century. Right: Luca Signorelli (–1523), *The Assumption of the Virgin with Saints Michael and Benedict* (detail), ca. 1493–1496. Metropolitan Museum of Art, accession number 29.164.

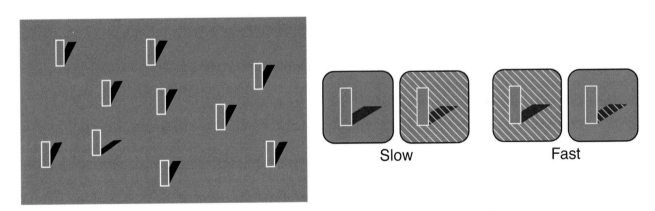

Figure 6.24

Visual search for an odd orientation is slow when the background texture crosses through the dark area as it would for a real shadow, creating X-junctions with the shadow border. But search is fast if the texture stops at the border, creating T-junctions, as if the region were opaque, or if the texture is only present inside. Both cases generate a scene border that aligns with the candidate shadow border, vetoing it as a shadow and making its properties, like orientation, easier to access. Image credit: PC, adapted from Rensink and Cavanagh (2004).

Figure 6.25

At the top of the picture, the shadow is aligned with a structural detail of the building: the shadow character is disrupted, and the column appears to be a different color from the rest of the wall. At the bottom, some cues support the shadow interpretation (i.e., the crossing with horizontal contours) and restore the shadow character. Image credit: RC, Yale, November 2013.

The noncoincidence rule has consequences in ecological scenes. The alignment of a shadow border and architectural features can be mistaken for a reflectance change, as in figure 6.25. Moreover, shadow disruption can be particularly compelling when the presence of a shadow would change the apparent lightness of surfaces. For example, the base of the glass in figure 6.26 appears to have a brownish rim in the picture. The glass is, in fact, colorless; placed directly under a source of light, it casts a self-shadow that encompasses the base, so that the shadow contour is aligned with the edge of the base, an alignment that vetoes a shadow interpretation.

Figure 6.26
A glass under light from straight above. The rim of the base of the glass looks brown-ish but is actually a shaded area that coincides with the glass's base. Image credit: RC, Hanover, New Hampshire, 2014.

Context can therefore determine shadow. This theme is larger than can be discussed in this chapter, and it opens the way to a discussion of the complex issue of inference and cognitive penetration. We return to the issue of top-down influences in chapter 7.

6.2.3 Shadows Must Be Flat

Shadows are a product of illumination and thus are properties of the surface they fall on. Dark areas that appear to lie flat make excellent candidates for

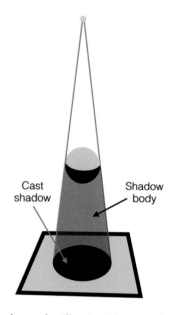

Cast shadow

Shadow body

A volume of unilluminated space exists between the object that casts the shadow and the surface on which it falls. This is the shadow body, which does have volume. The property of having no volume applies only to the cast shadow, not to the shadow body itself.

Figure 6.27
If the cast shadow appears to have volume, it loses its shadow label and becomes an object on its own. *Left*: A cast shadow has its own internal attached shadow, implying some round volume. *Right*: The shoe is actually a cast shadow of all the small wires, arranged to produce an image with the appearance of 3-D structure and reflections, completely losing shadow character. Image credits: Left: Hunting scene, Piazza Armerina, Villa Romana del Casale, fourth century. Right: Larry Kagan.

shadows, even if they are not. Critically, a shadow cast on a surface should not appear to have any volume of its own. Several late Roman mosaics and some modern paintings make the mistake (fig. 6.27, left), or the deliberate move, of giving shadows some volume—which takes away both shadow labeling and character. Some artists exploit this property to make actual shadows appear to be objects (as Larry Kagan does with the sculpture in the opening picture of this chapter and on the right in fig. 6.27).

A shadow cannot have volume of its own.

6.2.4 Shadow Shape Must Be Approximately Conformal

The relation between the shadow shape and the shape of the object that casts it is sometimes critical and sometimes not for shadow ownership, as we saw in chapter 5; but it is also important for labeling a shadow as a shadow. On top of blocking the shadow ownership, a mismatch between the shadow and object shapes may veto the shadow labeling.

The importance of the geometry of the connection between the shadow and the object was examined in detail in chapter 5. We saw that shape requirements are not coded precisely, only that:

Shadow shape must be approximately conformal.

This was shown for the case of anchoring, semidetached, and detached shadows (sec. 5.2.3). As a matter of fact, to ascertain whether shadow shapes conformed to the object, we relied on the subjective impression of perceiving or not perceiving a shadow—a consequence of labeling. Thus shape conformity traces a two-way path between ownership and labeling. This interdependency was clear in figure 5.11, shown again in the margin here, where the floating or nonfloating appearance of the cube was associated with the dark areas appearing as shadow or surface markings, respectively.

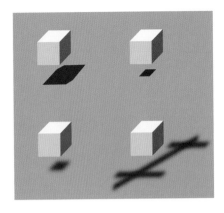

When the dark area is taken as a shadow, the object floats above the surface. Otherwise the dark area appears as a surface marking, and the object sits flush with the surface. Image credit: PC.

Figure 6.28
Unlikely lighting conditions. The shadow surrounds the object, and the symmetry or the unusual location may veto shadow labeling. Image credit: RC, Lefkada, July 2013, nighttime, artificial light.

Here is a further example showing that an unlikely shadow shape and location may also demote a shadow into a surface marking even if it is a real shadow. The shadow of the shrub in figure 6.28 looks more like a water stain or dirt spreading from the container, at least for some observers. What would explain this loss of shadow labeling and character for a perfectly accurate shadow? Perhaps the excessive symmetry. The sun is usually at some angle between straight overhead and the horizon, except at the relevant times in the tropics, and so shadows typically fall to the side of the objects that cast them. In figure 6.28, an artificial light source straight overhead generates an atypical, symmetrical shadow that surrounds the object. The real shadow looks less shadowlike, even in the presence of other cues (the outline of the shadow matches the texture of the shrub) that should favor a shadow interpretation. This suggests that:

The characteristics required for shadow labeling (and then ownership) must come from a range of expected or generic shapes and not from a computation of an accurate shadow.

Even if the shadow contours meet the object contours appropriately, the shadow shape must be qualitatively matched to the object shape. Figure 6.29 ("Fast") shows three examples of shadows behind objects that could

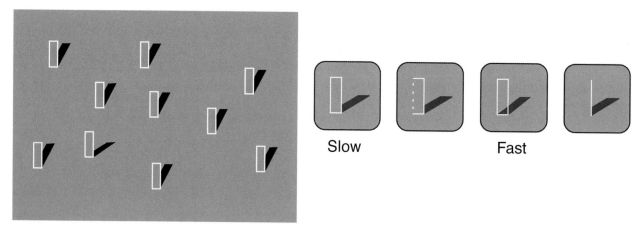

Figure 6.29
Finding the odd orientation is slow when the region adjacent to the bar is appropriate for a shadow: the shadows are demoted to nonsalient, accidental features. However, when the object shape does not conform to the shadow shape, search is fast, indicating that otherwise acceptable shadow regions have been rejected because of this mismatch and so remain as salient, easy-to-find scene features. PC, adapted from Rensink and Cavanagh (2004).

not have cast them, as the objects either are not bounded or are transparent. This mismatch in shape is enough to reject the dark areas as shadows and make their orientation more easily accessible in a visual search. The three examples tested here do not allow us to say what properties of shape mismatch are critical; they only allow us to say that an objective test shows that some types of shape mismatch veto the shadow label.

6.3 Shadow Properties That Are Not Mandatory

If the "darker along the border property" is mandatory, some other properties that are always true of physical shadows can be disregarded without disrupting shadow labeling and character. Here we discuss some cases of visual tolerance.

6.3.1 Color Does Not Matter

For the colors of shadows, or more specifically, their hue, we can identify two separate outcomes. First, the bluishness of shadows in outdoor settings leads to a bias to attribute bluishness to the illumination as opposed to the surface material; second, the actual physical constraints that limit the relations between hues inside and outside a shadow are ignored by our visual system.

Cast shadows are typically bluish outdoors: when an object has blocked the sunlight, the light falling on the shadow region comes mostly from the blue sky (along with some sunlight reflections from nearby surfaces). This common property of outdoor scenes is captured in many paintings and photographs, and humans seem to have adapted to it as we did with the light-from-above bias (chap. 4). As it happens, we often interpret bluishness in a scene to be a property of the lighting and not of the surface material.

In ecological situations, colored shadows arise when there are two light sources (with the sun and the blue sky as a particular instance), one that is blocked by an object, casting a shadow, and a second source that falls both on the shadow and outside it as well. That both light sources must sum in the surround of the shadow constrains the possible color in the surround: for example, a saturated green shadow can never have a saturated red surround. If the direct illumination is pure red, it must add with the ambient green, seen in the shadow but falling everywhere, and the result would be a desaturated, more yellowish red.

Blue snow shadows. San Secondo di Pinerolo, Italy, December 3, 2017. Image credit: Emanuela Zilio.

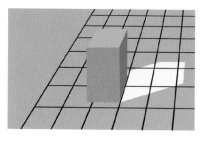 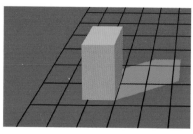 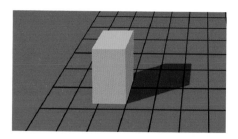

Figure 6.30
What matters for shadowness is that the shadow area be darker. Achromatic areas (*left*) or chromatic areas (*middle*) that are lighter than their surrounds veto a shadow interpretation. Impossible colors, where the surround does not include the light falling in the shadow, do not veto the shadow (*right*), provided the relevant area is darker. Image credit: PC.

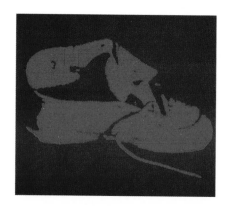

These properties are always true in the real world, but shadows can apparently take any color (fig. 6.30, right; fig. 6.31), provided that the shadow area is darker than its adjoining area. Objective tests with images like the two-tone shoe in the margin show that the contrast necessary to see the three-dimensional structure is the same for achromatic versions (light and dark gray) and for versions with impossible colors in the shadow area. This is odd because a color change at a border would seem a reliable indicator of a change in material rather than illumination. However, there can be color changes at shadow borders when there are multiple light sources: for example, shadows cast by direct sunlight are bluer than their surround. Nevertheless, color is often used in computer vision to distinguish material borders from shadow borders although it has little or no effect on human perception of shadows. We may hypothesize either that the physical rules for color are too complex to encode into neural mechanisms or that the frequency with which the color of dark areas would be useful in identifying shadows is too low to merit explicit verification. It may also be the case that the areas of the visual system that identify shadows process only luminance information.

Shadows may have impossible colors.

6.3.2 Conflicts of Illumination Direction

It is often the case that when there is one shadow, there are several. A scene may contain many objects—trees, people, rocks, buildings—each of which casts its own shadow, or even if the scene contains a single object, it may, if it is a bit complex, cast several shadows onto itself or on visually separated

Figure 6.31
The shadows of the trees have an impossible color but appear to be shadows none-theless. Georges Braque (1882–1963), *Paysage à l'Estaque (III)*, 1906. Art Institute of Chicago. Image credit: © 2015 Artists Rights Society (ARS), New York / ADAGP, Paris.

surfaces. In the natural world, all these shadows would typically have a single illumination source, the sun (or the moon, or perhaps a flash of light-ning, or Venus on a clear moonless night). As a result, all these shadows should point back to the same source of light. Is the visual system sensitive to consistent lighting direction across shadows? We can determine the illu-mination direction for local image regions when asked, although this may be more of a cognitive game that we can play when we line up an object and a shadow. Experiments that evaluate whether we are aware of light's direc-tion through empty space have not supported this notion, so it is possible that the early visual system does not extract geometric properties of the illu-mination but only gauges its strength and color once it lands on a surface.

In fact, it appears that we are extremely bad at noticing discrepancies in illumination direction. Figure 6.32 shows that we do not notice a striking inconsistency in the shadows, here on the front arch of the Taj Mahal. On the other hand, we do catch occasional illumination anomalies in some old movies or poorly doctored images (fig. 6.33). In these cases, it may be not only the mismatched direction or absence of shadows that we notice but also a mismatch of intensity or sharpness.

Figure 6.32

The shadow directions in this modified image have inconsistencies that are quite hard to notice. The cast shadow on the front arch of the Taj Mahal is for light from the right, but all other cast and attached shadows are for light from the left. Image credit: Muhammad Mahdi Karim; modified image, PC.

Figure 6.33

Floating civil servants. When the county mayor and associates went to inspect the newly constructed country road at Lihong Town, they forgot their cast shadows around their feet and ended up magically floating. Photo credit: Huili County Government website, June 26, 2012. Accessed March 2017.

Odd illumination directions have been tested in visual search tasks (fig. 6.34, left), and in some cases it turns out that we are quite sensitive to them. However, in these tests, both lighting direction and 3-D orientation were varied; the target of the visual search was the only item with an odd 3-D orientation. With common orientations but one different illumination, as in the example in the margin, the odd item may seem harder to locate, but it may be found based on its local features rather than illumination (it is the only item that is dark on its left side). For a robust test of illumination, the target is defined solely by its odd lighting direction, and the 3-D orientations of all items are random. In this case, the odd lighting direction is extremely difficult to find (fig. 6.34, right), indicating that lighting direction is not a feature that is easily accessed.

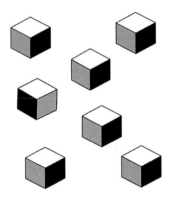

Why would our visual system not check for lighting consistency? Our indifference may derive from our evolutionary history, where a single light source, the sun, dominated and automatically enforced global consistency of shadows. We cannot easily appeal to the existence of multiple light sources. Although these are common now, they are too recent to have exerted any selective pressure. Nevertheless, we cannot rule out the possibility that we acquired a tolerance to multiple light sources in our early visual environment. Whatever the case, the human visual system may have

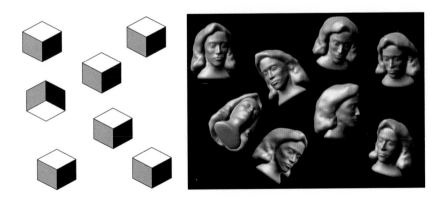

Figure 6.34
Left: When one of a set of identical items changes orientation and lighting, it is easily noticed, but this may be a result of the orientation change. *Right*: When lighting direction is the unique change on one object among several having varied orientations, it can take several seconds to locate (it is the one on the bottom left). Image credits: Ostrovsky, Sinha, and Cavanagh (2005).

little to gain by checking for global illumination consistency, since it is so widely true in natural scenes. More to the point for computational cost, local analysis typically works well enough for shape recovery and shadow labeling.

We are not always poor at noticing inconsistencies; they may become salient when they exist within a single object (or object part). Figure 6.35 shows that a reversal of lighting direction between a man's head and shoulder is not very noticeable, whereas the isolated reversal of the shadow on his nose is quite disturbing. So it may be after all that lighting direction is verified within a single object. However, this appears to happen only for familiar objects and may therefore just reflect knowledge of familiar patterns of shadows and not a true verification of common lighting direction. The reader may verify the reduction in the visibility of the shadow errors by turning figure 6.35 upside down.

In conclusion, we are remarkably insensitive to inconsistencies of illumination direction in experimental and natural settings. This suggests that the visual system does not attempt to verify the global illumination direction across several shadows.

Shadows do not have to conform to consistent directions of illumination.

Figure 6.35

Left: The head and shoulders are both illuminated from the right. *Middle*: The whole head has been reversed, so that it is illuminated from the left, while the shoulders are still illuminated from the right. This inconsistency is not too disturbing, although it becomes evident with some inspection. *Right*: In contrast, only the shadow on the nose has been reversed, and this jumps out as a somewhat grotesque disfigurement. Image credit: PC, adapted from Ostrovsky, Sinha, and Cavanagh (2005).

6.3.3 Illumination Conflicts: Cast versus Attached versus Shading

We have described the properties implicitly used by the visual system for labeling shadows (i.e., telling shadows from nonshadows.) As we have noted, shadows come in different types: cast and attached shadows. In ecological settings, the direction of illumination is perforce consistent across the two types of shadow. In addition, shading gives a third cue to light direction. The point here is that we find little or no evidence that the visual system verifies the consistency of lighting direction across the different types of shadow and shading, or even whether all three types are appropriately present together.

In the presence of a single light source, an opaque object casting a shadow has shading on its side that faces the light, and an attached shadow on the side opposite the light. This is an ecologically robust fact, endlessly confirmed in observation. However, dissociations of cast shadows from attached shadows are widely tolerated in perception. Configurations with cast but no attached shadows, and with attached but no cast shadows, are perceived as not particularly problematic. Without the cast shadows, we see the same positional uncertainty that is present with no shadows of any kind. This case of attached shadows without cast shadows is quite common in works of art during certain periods and provides the most striking example of the dissociation we have in mind. Interestingly, the use of cast shadows without attached shadows is—to our knowledge—nowhere to be found in the history of pictorial representation.

In Pesellino's *Construction of the Temple of Jerusalem* (fig. 6.37), we see an extremely detailed set of attached shadows (and possibly shading) in walls

Recall that shading is the variation in light returning from a surface as the surface changes orientation relative to the light. In the case of shading, the surface remains in direct illumination; if it curves away enough from the light source, then it then turns into an attached shadow.

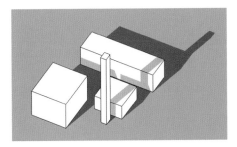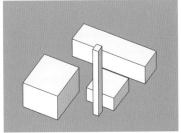

Figure 6.36
Cast shadows without attached (*left*) versus attached without cast shadows (*right*). Image credit: RC.

Figure 6.37
Attached shadows in the absence of cast shadows were the norm in Western painting
for many centuries. The attached shadows on most of the left-facing surfaces indicate
a light source to the right. This light should cast shadows of the beams, people, and
structures toward the left, but none of these are present. Pesellino (Francesco di Stefano) (1422–1457), *The Construction of the Temple of Jerusalem*, c. 1445. Tempera on
panel. Image credit: Harvard Art Museums, 1916.495.

and bricks, but no cast shadows appear in the painting. This dissociation
can be observed in virtually all representations before the Renaissance, and
in many during the Renaissance.

Can we identify an ecological counterpart for the attached-shadows-
with-no-cast-shadows situation? In the natural world, an overcast sky may
provide relatively evident shading and attached shadows coupled with
weak cast shadows, but the complete dissociation of the two is not possible in natural scenes (although it could be contrived with advanced lighting techniques). That we do not find representations such as Pesellino's

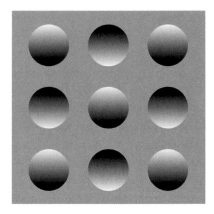 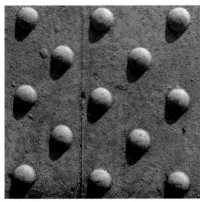

Figure 6.38

Left: Typical convex and concave surfaces created by linear shading gradient for research on depth from shading. *Right*: The shading, cast, and attached shadows of actual convex bumps protruding from a surface. Image credit: RC, February 2014.

particularly disturbing suggests that we analyze cast and attached shadows and shading independently and that there is little or no cross validation of the recovery of illumination direction, if it is actually recovered at all.

We find a similar example of shading and attached shadow without cast shadows in the modern literature on shape from shading (fig. 6.38, left), where we might expect that the absence of valid cast shadows would at least have been mentioned. Here, if the round items are seen as convex because of their shading (or attached shadow), then they must also cast a shadow on the adjacent surface, as can be checked in real-life situations such as the right-hand image in figure 6.38. But we are not disturbed by the absence of shadows, and it is striking that this tolerance for a lighting error has only recently been mentioned in this area of shadow research.

Finally, and even more interestingly, our visual system tolerates a wide range of impossible arrangements of cast shadows. In particular, "carpet shadows" (see chap. 10) run along the ground but do not climb the intervening objects' surfaces as they should. In figure 6.39, on the left, carpet shadows are paired with the absence of attached shadows; on the right, carpet shadows are paired with attached shadows.

Cast shadows, attached shadows, and shading can appear independently.

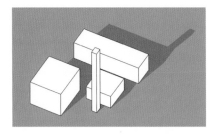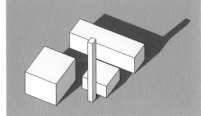

Figure 6.39
Carpet shadows, without or with attached shadows. Image credit: RC.

6.3.4 Mismatch of Casting Object and Shadow Shape

Although some checks on shadow shape exist (as described in sec. 6.2.4), impossible shadows are common in paintings and are seldom noticed (see fig. 6.40 for an extreme example). This tolerance for deviation is striking enough to merit an entry among shadow properties that can be broken. Why are we so tolerant of shape deviations? Many shadows are created by objects outside the field of view; their shape may be distorted by the surface relief; their size depends on whether the light rays are parallel (the sun) or divergent (a nearby artificial source) and whether the source is small or extended. It is not unreasonable that the visual system does not (and often could not) require the precisely correct shape and can get by with a few shape properties that are easier to deal with, as outlined in section 6.2.4. This broad indifference to shadow shape is important for those working on rendering, as it allows huge savings. If vision does not care too much about shape and position, why should a visual artist?

The tolerance for incorrect shadow shapes makes a fascinating subject for empirical investigation: how much distortion can the visual system tolerate and still treat a given area as a shadow? Tolerance can be surprisingly high in certain contexts. In figure 6.41, the shadowlike areas on the ground are in fact areas of different reflectance (the pavement consists of both light and dark gray stones). Although the dark regions do not match the shapes of the empty boards, they preserve properties like angularity and elongation that would be expected in their shadows.

Impossible shadow shape does not always veto a shadow.

Figure 6.40
The shadow of the key is convincing despite having no correspondence to the object
that casts it: a hook with no key on it. Image credit: Jiro Takamatsu, *No. 277*. From
the Minnie B. Odoroff Collection. Photo by Bruce Goldstein.

6.3.5 Shadows Do Not Have to Have Specific Border Profiles

Natural shadows have borders with a gradient from dark to light, but that
gradient, the penumbra, can have almost any width. The border can be
sharp if an object casts a shadow on a nearby surface or if the illumination
is a point source, such as a reflection from a curved surface. Conversely, the
border can be extremely broad with a shallow luminance gradient if the
shadow falls on a distant surface or the light source is diffuse. Many texts
and introductory courses in perception or art suggest that the blurry edge of
a shadow, the penumbra, is a defining characteristic. In fact, a sharp boundary does not rule out an area as a shadow for a human observer. Blurriness

Figure 6.41
Fake shadows on a street. Image credit: RC, Copacabana, April 2013.

on its own is not an obligatory property, but in some cases, lighting conditions such as an overcast sky or large separation between the object and its shadow do specify a blurred penumbra. In such cases, human observers can ignore the required blurriness and accept physically impossible, sharp shadows. It appears that our visual system does not check the consistency between the properties of the light source, the distance between the object and its shadow, and the width of the shadow's penumbra, or even the similarity of shadow properties across different shadows (fig. 6.42). Once more, this would appear reasonable given the computational cost of such a consistency check. Nevertheless, human vision may have a perceptual bias in favor of the presence of a penumbra as a result of its frequency in natural scenes. We would need to test cases of ambiguous shadows to see if blurred borders favor a shadow interpretation. This would be similar to the bias for lighting from above that affects convex versus concave depth from shadows for otherwise ambiguous stimuli.

   Similarly, cast shadows do more frequently have straight or straightish boundaries because they are the projections of the object's silhouette,

Figure 6.42
The man's shadows have been digitally manipulated to exchange positions in these two images. *Left*: Under cloudy skies, we should see only minimal shadow, but the presence of the sharp shadow from the sunny image does not seem out of place. *Right*: Similarly, the direct sun should cast a strong shadow, as it does for other objects in the scene. Nevertheless, the presence of only diffuse shadows near the feet seems acceptable. Image credit: PC, adapted from Brigitte Cavanagh, Vincennes, September 10, 2017.

which in general has fewer concavities than any single section through the object. A complicated edge falling on a smooth surface is therefore less likely to be a shadow. Again, this is not an obligatory property of shadows. Complicated edges do not veto shadow interpretations if this is still the most likely interpretation, but only bias it in cases of ambiguity. The dominant factor that tips the interpretation one way or the other is not the complexity of the cast shape but often the "goodness" of the three-dimensional object that may be supported by the shadows (e.g., the face in the margin of the previous page). Conversely, if the cast shadow depicts a "good" object, this alternative may veto the shadow interpretation (the shadow that looks like a shoe in the margin).

In summary, these two border properties—penumbra and straightness—are not obligatory physical properties of shadows. They may nevertheless bias shadow interpretations when the situation is ambiguous, but otherwise they can be broken without vetoing shadow labeling.

The profile of a shadow border may bias but not veto the acceptability of a region as a shadow.

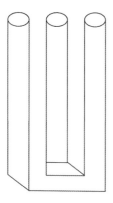

Devil's fork. Image credit: RC.

6.3.6 Shadow Character Is Not Monolithic

Once a part of a dark region is labeled as a shadow, we might expect the label to propagate throughout the entire region, but this may not always occur. In particular, it is possible to see a uniformly dark region as a shadow in one area but a dark surface in another (fig. 6.43).

The situation is similar to the changing interpretation of the tines in the devil's pitchfork. A shading analogy of the devil's fork has been provided by Edward Adelson (fig. 6.44), where the light and dark stripes change from being the shading cues of a beveled edge to being light and dark paint.

The point of these changing assignments is that shadow labeling (like the object structure of the devil's fork, or like the shading in fig. 6.43) is not enforced throughout a contiguous area but can change within a connected region. A similar point is made for repeated shapes that can be seen as shadows in one context and water stains in another, even though they have the identical shape and luminance (fig. 6.45).

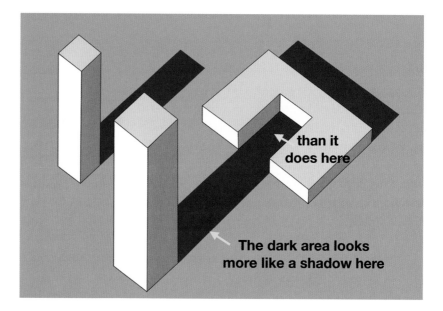

Figure 6.43
A region that is half shadow, half dark material. The shadow interpretation is consistent with the illumination of the blocks. But as the shadow boundary ends up coinciding with a surface discontinuity (at the C-shaped block), the dark region now appears to be a dark material. Image credit: PC and RC.

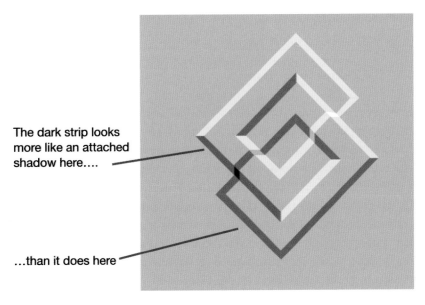

The dark strip looks more like an attached shadow here....

...than it does here

Figure 6.44
The same shade of gray is read differently in different portions of the picture. Image credit: PC, adapted from Edward Adelson.

Figure 6.45
Context can affect shadow labeling. Image credit: RC, redrawn from an idea of Jayme Jacobson.

A uniform region may be labeled as dark shadow in one portion but dark material in another.

6.4   The Shadow Merge Test

Here, we present a do-it-yourself test for shadow character. If you doubt the nature of a dark region in the environment (is it a shadow or a dark material fully illuminated?), you may try the shadow merge test. Position another object to cast a new shadow on top of the region in question. If the region is a shadow, it will merge with the new shadow, without any contrast between the two dark regions (fig. 6.46). If the region is a dark material, however, the new shadow will continue into the darker region. The shadow merge test is not foolproof, for instance if the object casting the second shadow is much closer to the receiving surface its shadow will still be visible within the shadow of the more distant object (fig. 6.47).

6.5   Configurations That Use Shadow Rules to Make "Faux" Shadows

Light comes from above, but so does dust. Or, more to the point, dust accumulates on the top surfaces of objects. If the color of the objects is lighter than that of the dust, the dusty area has the appearance of a shadow, and the object itself may appear to be lit from below.

These days, many buildings are lit from below at night, as it is much easier to put lamps on the ground than high up above the buildings. This

Figure 6.46
Is it a shadow? The shadow merge test will tell. Image credit: RC, Paris, July 2015.

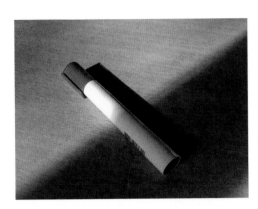 

Figure 6.47

The shadow merge test may fail if the surface is too close to the test object. In that case, the object will screen enough ambient light from the environment to enhance the shadow contrast. *Left*: A single source of light casts both the large shadow and the shadow of the marker. *Right*: This shadow within a shadow provides a rough heuristics to fix the penumbra problem that plagues large sundials discussed in section 1.2.5. A twig inserted in the penumbra is moved until its shadow disappears, signaling the approximate edge of the penumbra where no part of the sun is visible and, consequently, making it possible to read the time indication on the dial marked in steps of two seconds. Image credits: Left: RC, Paris, February 2015. Right: Barry Perlus, www.jantarmantar.org.

placement is often criticized, as it contributes to light pollution, as well as giving buildings an unnatural, spooky appearance. Dusty accumulations can also give the impression that a structure is lit from below (fig. 6.48), albeit without the light pollution issue.

Other situations mimic shadow patterns with distributions of dark and light on surfaces. In figure 6.49, a light fixture on the wall protects a region below from falling snow, creating an area that the visual system is ready to interpret as a shadow. These *visual* faux shadows are actually analogues of shadows in the sense defined in chapter 7, but they are not holes in light.

6.6   Conclusions

In this chapter, we have studied the many conditions under which an area in the visual field appears or fails to be labeled as a shadow by the visual system. We have presented the following set of properties that must be present for successful shadow labeling by human observers:

Figure 6.48
Faux shadows created by dust. Paris, Charléty Stadium, March 2015. Image credit: RC.

Figure 6.49
Faux shadows created by snow. Image credit: RC, Bardonecchia, February 2015.

Shadows must be darker than their surround along the border. This is the basic
   criterion for shadow labeling.

The ratio of luminances is consistent along the edge.

The shadow label is propagated into the region bounded by the shadow border.

Shadows must cross surface contours, not coincide with them. X-junctions where
   shadow contours cross surface markings provide strong evidence for a
   shadow region. In contrast, any alignment of the shadow border with
   surface borders makes the luminance change more likely owing to a
   difference in materials, not a difference in illumination.

A shadow cannot have volume of its own. The candidate shadow region must
   lie flat on the receiving surface. It cannot look independently like a
   volumetric object.

Shadow shape must be approximately conformal. The restrictions on shadow
   shape are approximate except for the requirement that an anchoring
   shadow must meet the object contours that cast it. Otherwise there
   is some tolerance for deviations of shadow shape from the physically
   correct shape as long as the deviation is not extreme: the characteristics required for shadow labeling *must come from a range of expected
   or generic shapes* and *not from a computation of an accurate shadow.*
   The limits of this visual tolerance have not yet been appropriately
   tested.

The limited number of these obligatory properties makes the prediction of
illusions possible: some actual shadows are not seen as shadows, and conversely some regions that are not shadows are seen as shadows. For instance,
it is possible to fool the visual system into believing that there is a shadow
in a certain spot, by manipulating the reflectance features of the spot. This
is the reason we can perceive dark regions as shadows in photographs and
paintings despite violations of the natural properties of shadows.

   The following is the set of ecological and geometric shadow properties
that can be violated without destroying the shadow labeling that allows
human observers to recover scene structure and depth.

Shadows do not have to conform to a consistent direction of illumination. When
   several cast shadows are present, their locations do not have to be consistent with a single light source. This allows artists to take great liberty
   with the placement of shadows, so as to avoid situating key elements

of the scene in low illumination. There is a possible restriction where consistent lighting may be required within a single, familiar object.

Cast shadows, attached shadows, and shading can appear independently. In a natural scene, all these illumination effects should appear together. Nevertheless there seems to be no check on the consistency across these three types, and each can show up on its own or with others with little impact on the acceptability of each as an illumination effect.

Shadows may have impossible colors. The surround must always include the light falling in the shadow: a saturated red shadow with a saturated green surround could never occur naturally. Nevertheless, this inconsistency is ignored and supports the recovery of depth from the darker region.

Impossible shadow shape does not always veto a shadow. As mentioned above, it is helpful to have a shadow shape that is conforms approximately to the expected shape, but violations can be tolerated.

The profile of a shadow border may bias but not veto the acceptability of a region as a shadow. The shadow border can be sharp; it does not have to be a blurred penumbra. It can be complex; it does not have to be straightish or smooth, as ecological shadow borders tend to be.

A uniform region may be labeled as dark shadow in one portion but dark material in another. Although shadow labeling should propagate in from the shadow border to fill the contained region, it can stop when it meets a conflicting assignment from adjacent borders. The labelling tolerates nonmonolithic regions.

Thus our visual system actually uses only a few of the physical properties of shadows to determine if a region is or is not a shadow. The two main rules that cannot be broken—*darker along the shadow border* and *no extra contour alignment along the border*—are properties that can be evaluated locally and thus are quickly verified and most often sufficient. Here, as in many other cases, the visual system proves to be parsimonious and tolerant to speed up processing.

Having defined the contours of the shadow mission, and the conditions that make it possible, we turn now to a different set of topics. How deep do rules for shadow labeling cut into cognition? Do shadows have a life beyond the visual system?

Further Research Questions

1. Is a shadow interpretation favored when blurred borders are present in the image?

2. In addition to those discussed in this chapter, what other conditions disrupt shadow labeling and character? Do context and object knowledge play any role, and how?

3. Can we manipulate shadow perception in ecological settings? Based on what we have learned so far, one possibility is to turn X-junctions into T-junctions by exactly superposing a shadow on a figure that has a different reflectance than its background.

Figure 6.50
Grains of rice fill precisely the shadow of the stovetop coffee maker. The rice has a higher reflectance than the floor so ambient light makes the shadow of the wooden spoon darker than the shadow of the coffee pot. The shadow of the spoon therefore appears to pass under the grains. Image credit: RC.

The light–shadow boundary: are these dark regions shadows, or are they figure, or background? Image credit: RC, Paris, July 2017.

7 The Shadow Concept

7.1 What Is a Shadow? We Are of Two Minds

Tourist guides to Piazza Armerina's mosaics in eastern Sicily routinely claim that some of the images provide evidence for the Mediterranean invention of skiing (fig. 7.1).

As Mount Etna is not far away, its top covered with snow, this historical reconstruction may sound plausible. However, the object-like structures on which some of the characters in the mosaics stand are not skis, and not even solid objects at all, but poorly represented, highly conventionalized shadows. This image makes the point that both perception (what we see) and cognition (what we think about what we see) have expectations and rules about what is a shadow. In figure 7.1, for us, both systems agree that these "skis" are not shadowlike. Only people from the late Roman era would have thought, "Of course those are meant to be shadows." Nevertheless, we can be fairly sure that the spatial layout a Roman observer saw in the image would not have been influenced by these fake shadows. A Roman's perception would have been unaffected by the knowledge that the "skis" are meant to be shadows.

We get two opinions on shadows: one is just what we see, and the other is what we reason about what we see. Sometimes they agree; sometimes they do not. As we have seen, perception has powerful computational and conceptual resources for making inferences about scene layout and illumination based on shadow information. This level of decision making has been called *visual cognition* or *unconscious inference*. It stands apart from our general conscious, cognitive processing, where we can reason about scenes, goals, and, at times, shadows. Cognition mostly takes the perceptual descriptions and uses them to make further choices. We may choose to

Figure 7.1
The so-called skier mosaic at Villa Romana del Casale, Piazza Armerina (ca. 300–320).
Although often interpreted as skis, the dark regions around the feet are just the con-
ventional, and very ineffective, representations of shadows that were the required
style at the time. Image credit: Kenton Greening.

head toward a shadow for the coolness it offers, for example. When cognition bothers to check perceptual representations, it most often agrees, but
when it disagrees, it rarely alters our perceptual experience. Realizing that
the artist intended the "skis" in figure 7.1 to be shadows does not make
them look so. This is the "cognitive impenetrability" that Zenon Pylyshyn
described as the barrier between cognition and perception. Although we
cannot "correct" the perceptual experience, we may notice that our perception has made a mistake or an interesting choice, thereby revealing some
properties both of the underlying perceptual architecture and of our cognitive concepts concerning shadows.

Here we probe the conceptual architecture for shadows at both the perceptual and the cognitive levels by putting them under some pressure. We look at cases where perception makes errors that can be noticed by cognition—extensions of our shadow disruption tests in the previous chapter. We also examine cases where perception is fine but cognition is puzzled (is that really a shadow?), and one example where the cognitive conceptual structure for shadows is plainly wrong but perception is unfazed.

7.2 Perceptual Concepts of Shadows Gone Wrong

In figure 7.2, it looks as if something is missing in the shadow of the racquet strings. The frame and the strings of the badminton racquet cast their shadows on the surface underneath. But the gray logo does not. The physical explanation is obvious: the white parts of the strings block the light as well as the gray segments, so the shadow is unaffected by the color of the strings. However, owing to perceptual color spreading, the gray region appears to fill the area defined by the dark strings, including the holes between cords: an area in the shape of the logo. Now, if the filled shape of the logo were a physical object, it would block light and cast a filled shadow (an area in

Figure 7.2
The gray logo is painted on the cords of the badminton racquet, but it looks as if something is missing in the shadow. Image credit: RC, Paris, September 2012.

Shadows spread over empty areas of the fence. The filled-in areas appear to cast a shadow on the ground visible in the recess behind the grid. Image credit: RC, Pisa, June 2010.

the shape of the logo, completely darkened). That we are surprised by the absence of the shadow indicates that the perceptual system believes that the filled gray logo has object status and should be opaque to light, and this calls for the presence of a shadow. The cognitive system can reason about the surprise absence and is briefly amused by the error.

Whatever looks like an opaque object is expected to block light and to cast shadows.

In figure 7.3, we see a round-shaped shadow on the wall. We trace it back to the nearest figure that may correspond to the object casting it. That figure turns out to be the shadow on the chair, which again through color spreading appears to be an opaque thing.

Thus perception accepts that a shadow can cast a shadow. Cognitively, we can discover the misattribution, but perception is unfazed.

Even an illusory object generated by a shadow is expected to cast a shadow.

These two examples show cases where expectations bear witness to some of the architecture of the perceptual processing of shadows, making errors that cognition can reveal with further reasoning.

Figure 7.3
A shadow of a shadow. The shadow on the wall grabs the first plausible figure that may cast it; this turns out to be a shadow. Image credit: RC, Paris, 2012.

7.3 Cognitive Concepts of Shadows Broader than Perceptual

7.3.1 A Visible Boundary Is Required for Shadow Labeling

Cognitively we can experience and complain about being in the shadow of a building even when no unshadowed areas are visible. This is common, for example, in much of Manhattan. Perception is more limited in its concept of a shadow. Without a border, a region that is blocked from a direct illuminant does not act as a shadow—take nighttime as another obvious example (fig. 7.4). Clearly, without the presence of light and dark areas and the illumination border that separates them, the issues of shadow labeling and recovering scene layout from shadows do not exist. These perceptual processes are border based.

For perception, a shadow must have a visible boundary.

Figure 7.4
The sun has set, and we are in the shade of the Earth. Nevertheless, we do not spontaneously perceive or conceptualize twilight or night as a kind of shade or shadow. Image credit: RC, Barbizon, August 2013.

Figure 7.5
Light enclosures. Image credit: RC, Florence, September 2009.

However, even when a border is present, the darker region may not be treated as a shadow if it dominates the scene (fig. 7.5). The visual system may deal more efficiently with bounded areas, and if the light regions are smaller, they become the conceptual focus.

We will see later (in chap. 8) that these light enclosures are the dual of shadows: they are local presences of light, enclosed within a shadow.

7.3.2 Perception Acts on Flat Shadows, Not on Shadow Bodies?

Are shadows flat entities or volumes? We typically reason about shadows as two-dimensional things. The shadow cast on a wall has no thickness: it is *flat*. The attached shadow on the unilluminated side of the object is a surface feature with no volume. But there is a volume of space that is blocked from the light source extending between the object that interrupts the light and the surface on which the resulting shadow falls. English has no specific name for the three-dimensional region of air or empty space that is shaded by an object. This region is visible only on occasion, generally because of the presence of dust or mist in the air. We use the term *shadow body* to describe the whole region of space where light is blocked. We can reason about shadow bodies, but they do not seem to enter into the perceptual processing of shadows. Perception acts on flat shadows.

The advantage of the *flat shadow* account is that it conforms to our naive concept of a shadow as figure. Shadows are visible on a surface (to which we normally point when asked about their location), and if they have a shape, this is the 2-D shape we usually ascribe to them. But shadows do not always have the integrity of ordinary shapes, and perception seems to make no attempt to piece the shadow parts together (sec. 4.5.3). When a railing projects a shadow that falls over two steps, it is split on two surfaces. Is this *the same* shadow? A shadow body lets us reason about this, even if we are not accustomed to reasoning about spaces instead of surfaces. The shadow

Split railing shadow. Image credit: RC, Ithaca, Greece, August 2017.

Figure 7.6
The visibility of shadow bodies. The shadow cast from Dubai's Burj Khalifa, as of today the tallest building in the world, seen from the ground. There is enough shadowed atmosphere between us and the top of the building to produce a darkened region against the background of the illuminated sky. The building's cast shadow falls on the ground (except at sunrise or sunset), outside of view, and we see the shadow body as continuing from the building. Note the grading of shadow intensities from different widths of the building. Image credit: Goffredo Puccetti, January 2017.

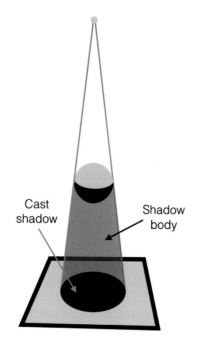

Cast shadow

Shadow body

A shadow body is bounded by the attached shadow on the object and by the cast shadow on the projection surface.

Recall again that shading is the variation in light returning from a surface as the surface changes orientation relative to the light. In the case of shading, the surface remains in direct illumination; if it curves away enough from the light source, it then turns into an attached shadow.

body stretches from the object's face that does not receive light (this corresponds to the *attached* shadow) to the cast shadow on the receiving surface or surfaces. The dark face of the object and the 2-D shadow on the receiving surface form two object boundaries of the shadow body; the remaining defining boundaries are set by light. The shadow body thus has an enveloping bright surround and two dark ends. So the two flat partial shadows of the railing projecting on the two steps are both boundaries of a single shadow body. As mentioned earlier, if any dust or mist were in the air, this darker shadow body would be visible and would join the split half shadows.

If no surface existed to receive the cast shadow, the shadow body would extend out like a dark beam. Some authors have argued that the visual system maintains an unconscious representation of the direction and intensity of the light in the scene, so that the presence or absence (within a shadow body) of an illuminant would be available for any point in question. This "estimated lighting model" is controversial, and we have no evidence in any case that we naturally use shadow bodies in perceptual processing unless they are visible, and even then they just act as three-dimensional scene structures without any specific reference to shadowness—they could just as well be jets of water or mist. Nevertheless, we can reason about shadow bodies conceptually even when there is no material to make them visible. Interestingly, when they are visible (e.g., due to atmospheric conditions), we do not consider the dark areas between visible light rays as shadow bodies unless they are partially or fully enclosed by the light rays. Possibly, if there is a competition between shadows and light, light wins by default.

We can reason about shadow bodies, but unless they are made visible by dust or fog, they are not used by perception.

Painters do not seem to have noticed shadow bodies, except in the representations of spectacular sunsets, where they are the complement of light rays. (An exception is the Turner reproduced in fig. 2.7.) Finally, technical jargon distinguishes between shading and attached shadow, but it is not clear that we have a mental category for attached shadows as different from shading. It is possible that perceptual and cognitive processes do not consider them as shadows at all, reserving the label for cast shadows. The attached shadow contour provides information about surface curvature that is closely tied to the shape recovery from shading flow (the lines of equal luminance over the object's illuminated surface). No more illumination change (due to the main light source) occurs beyond the edge of the attached shadow, and so no more information about the surface relief (other than that provided by any ambient lighting). The point is that the attached shadow does not participate in the perceptual recovery of the scene layout in the way the cast shadow does. On the other hand, the attached shadow is an integral part of the shadow body.

7.4 Cognitive Concepts Thwarted, Perception Unfazed

7.4.1 Filter Shadows

A colored object that is partially transparent can cast a colored shadow (fig. 7.7). Sorensen called these *filtows*, but we prefer the term *filter shadow*. These are intriguing because they bridge the difference between shadow and light. As with an ordinary shadow, the filtered light is darker than its surround and triggers the regular perceptual missions that shadows fulfill.

Figure 7.7
A filter shadow, the brown shadow of a brown bottle. Image credit: RC, Recloses, June 2007.

An ashtray filters some light. Whether the filtering is seen to be a property of the light source or a property of the shadow depends on the portion of the light that is filtered and on the visibility or invisibility of a border.

Nevertheless, we encounter difficulties in reasoning about a filter shadow: is it a colored shadow or a colored illuminant; is it a shadow at all?

For example, take a green filter, such as an ashtray made of green glass. Put it under a light source, at some distance from it; light travels through the glass and hits a tabletop. Is the green area over the table a shadow? In the limiting case, if the ashtray is placed so close that it filters all the light, then no illumination border is present, and the green becomes a property of the illuminant—the entire room is now lit by a green light. As in section 7.3.1, no border, no shadow. If the ashtray is gray, it is unlikely that you would say that there is a gray *light* filling the environment. We may be inclined to treat the effects of uncolored filters as normal light and shadow phenomena. A standard cast shadow is only the limiting case of complete filtering by an opaque object.

If the transparent, colored material is filling a hole in an otherwise opaque material (fig. 7.8), then it casts colored light spots (see chap. 8), lighter than the background and so not filter shadows.

A filter shadow acts as a shadow for perception but is ambiguous for cognition.

7.4.2 When a Proper Umbra Makes a Poor Cast Shadow

Shadows are visual shapes; they can be described in terms of their figural properties. The great interest in shadows as figures is testified by the development of technical language describing them. The same language may also, however, introduce complexities that challenge commonsense intuitions.

Reforms by technical language happened, for instance, in astronomy with the notions of *penumbra* and *umbra*. Given that most light sources are not point-like, each single point on their surface casts a shadow that differs slightly from, and can overlap with, shadows cast from any other point.

An *umbra* is the intersection of all point-generated shadows.

A *penumbra* is the union of these shadows minus their intersection, that is, minus the umbra.

These notions are not well aligned with the commonsense notions of cast shadow and penumbra: commonsense penumbrae are just the fuzzy boundaries of cast shadows, and cast shadows are in some way expected to resemble a silhouette of the object that casts them. In figure 7.9, the union of the various iterations of the shadow of the hand, in lighter and lighter

Figure 7.8
Light spots or filter shadows? It depends on the contrast with the surrounding area.
Image credit: RC, Paris, June 2007.

shades of gray, minus their intersection, satisfies the technical definition
of a penumbra but does not look like a commonsense penumbra. Neither
does the umbra of the hand, that is, the darkest area in the image, which is
the intersection of all the shadows of the hand from a number of different
sources, look like a commonsense shadow of the hand—it does not resemble a silhouette of a hand. Perception is not so fickle as our commonsense
cognitive notions about shadows. For example, the triangle shadows generated under cloudy skies (see chap. 9) where an object rests on the ground
offer an interesting and ubiquitous example of natural umbrae. They do
not resemble a cast silhouette of the object, but because they connect to

Figure 7.9
Light from multiple sources exemplifies the phenomena of penumbra and umbra. Technically, the umbra is the intersection of all the individual shadows, the darkest region, while the penumbra is what is left when you subtract the umbra from all the shadows. Image credit: RC, Paris, Charles de Gaulle Airport, September 2007.

the object, they serve quite well for perception to anchor the object to the ground.

7.5 Cognitive Missteps: Shadows as Things

The Yale Problem. In 1975, Todes and Daniels described a set of common-sense principles for reasoning about shadows, which they then claimed were self-contradictory. The self-contradiction reveals a basic problem with reasoning about shadows: we take them as things, whereas, in fact, they are nonthings. Here are Todes and Daniels's principles that appear at first glance to be reasonable, intuitive notions about shadows.

I.   If an opaque object casts a shadow, then some light is falling directly on it.

II.   An object cannot cast a shadow through another opaque object.

III. Each shadow is a shadow of some object.

However, these principles lead to a self-contradiction in a very simple case: Put a cube into the shade cast by a sheet of paper, then focus on the area where the cube would have cast its shadow if the paper were not there. If the paper is present, it blocks the light to the cube, and the area below the cube is still dark: so what is this area a shadow of? Not of the cube (principle I: the cube is not in the light). Not of the sheet (principle II: the sheet cannot cast a shadow through the cube). But since the cube and the sheet are the only candidates around, principle III would be false: there is a shadow that has no object casting it.

This inconsistency is not a problem of physics (the shadow does not vanish because of self-contradiction), so it must be a problem in how we reason about shadows. The problem is that we think of shadows as things, and so does the perceptual system, linking them to casting objects and deriving depth information. But, of course, a shadow is not a thing, and nonthings cannot be blocked: only light can be blocked. It is the paper that blocks the light, and the light is then absent from the volume defined by the light source and the paper. It makes no difference what is in that volume; it cannot further block the light.

The trouble with principle II is easy to grasp if we ask what the alternative would be. If a shadow cannot be cast through an opaque object, then what does come out the other side? Light? Clearly not. Does this mean we must accept that a shadow actually does pass through an opaque object, continuing on beyond it as further shadow? This proposal is technically correct—there is no light on the other side of the opaque object—but it is an uncomfortable one. The problem arises because shadows are not things that can be either projected or blocked. Only light can be projected and blocked.

The Yale Problem arises in other domains as cases of *causal preemption* that can underlie legal conflicts. If a neighbor's building would cast an illegal shadow on your house, can you stop the neighbor's construction even though a taller skyscraper blocks the light onto your neighbor? (Place your house in the dotted area of the problem.) The new building arguably does *not* cast a shadow on your house, and neither would the skyscraper once the new building is completed. Similarly, if you fire a handgun at someone who dies from natural causes just before the bullet arrives, are you guilty of murder?

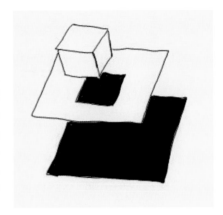

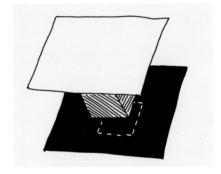

Top: Can an object cast a shadow through another object? The cube casts a shadow on the sheet, and the sheet casts a shadow on the ground. But we would not say that the cube casts a shadow on the ground. That shadow would have to "pass through" the sheet. *Bottom*: The Yale Problem. The cube is overshadowed by the sheet of paper. Consider the zone singled out with the dotted white line. Is it a shadow of the cube or of the sheet? Either answer challenges the commonsense account of shadows.

Although this sort of problem can lead to legal battles, it is helpful in the instance of shadows, as it reveals that the perceptual and cognitive systems have made an important blunder: shadows cannot logically be treated as things. Nevertheless, treating them thus serves both systems well. Shadows, when treated as things, allow rapid access to information about the location and shape of objects that block light from falling on other surfaces. The shadow is useful as a thing because it can be identified; if our visual and cognitive systems stuck to only tracking and objectifying light (as most computer rendering systems do), processing would probably be more cumbersome and slow. Developmental studies have shown that toddlers do treat cast shadows as objects, expecting them to move with the surfaces that they are cast on as if they were fixed surface features, and not with the objects that cast them. Despite this advantage for treating an absence of light as a presence of a shadow, we must keep in mind that our common-sense ideas of shadows may at times lead us astray.

Cognition may treat shadows as things that can be projected or blocked.

7.6   Perception Favors Object over Shadow Interpretations

Shadows, like objects, have shapes and sizes; they can move and, to an extent, be counted. Their reality is closer to that of material objects than to that of dreams or numbers. In some cases, shadows can be mistaken for objects, when their shape is more recognizable than is the object that casts them (fig. 7.11). While perception treats these "recognizable" shadows (sec. 10.2.3) as objects, cognition can realize that they are also shadows. Not only developmental studies suggest that infants initially treat shadows like objects; countless myths have provided shadows with a life of their own. Here we present some mechanics of the perception of shadows as objects.

All shadows are linked to an object that casts them, and so in some sense they "represent" that object, but some shadows also stand out as recognizable shapes on their own (*it's a man!*) or even identifiable shapes (*it's John!*). These are shadows that preserve information-rich profiles of an object, like silhouettes. Shadow theater representations make us falsely believe that, say, a dog is casting a shadow, when instead it is the actor's hands. This embodiment of the shadow's shape takes advantage of the compelling recognizability of some shadows. Most commonly, in natural situations, recognition-enhanced shadows are cast at specific times of the day:

Figure 7.10
Left: We recognize a frontal silhouette as a face but have difficulty identifying it as a particular person. *Right*: A profile silhouette makes it easier to identify its owner. Image credit: RC, Paris, September 2017.

for instance, for people standing, recognizable shadows appear at dawn or sunset on vertical surfaces, and all other cases produce inevitable and more or less damaging distortions and compressions; for flying birds, readable shadows are cast at midday on the ground; and so on.

Various myths about the origins of painting and visual representation in general draw on the recognition of shadows as objects. But which shadows elicit recognition? Shadows have shapes, and some of these shapes correspond to profiles of *canonical views* of objects. For instance, a canonical view of a human head is a frontal view, not a view from above. However, the silhouette of the canonical view of a human head, a frontal view shadow, is not easily readable (fig. 7.10). A canonical view of a horse is a side view. The silhouette of a side view of a horse is perfectly readable; it is thus a silhouette-optimal canonical view (fig. 7.11).

The canonical-view heuristic can explain why we perceive some shadows as objects (fig. 7.10). In discussing this phenomenon, Roy Sorensen suggests a simple heuristic for the promotion of the shadow and the demotion of the object, according to which "the more precise thing is the object and the coarser thing is the shadow." However, a simple control for this heuristic is an upside-down version of the horse picture, which weakens the assignment of object and shadow (fig. 7.11, right).

Figure 7.11

Left: The shadows of the toy horses are seen as objects, and the horses themselves are not easily recognizable. *Right*: With the image upside down, the horses are more object-like, and shadows more shadowlike. Image credit: RC, Barbizon, August 2016.

When a shadow projects a more canonical view than the object itself, the shadow may be perceived as an object.

Indeed, canonical views are orientation sensitive. An upside-down silhouette of a horse is less recognizable than a right-side-up silhouette. In figure 7.10, left, cast shadows are seen right side up, thus enhancing recognizability. Moreover, in that picture, light comes from below in the image—an uncommon condition—which means that the shadows that are actually seen are in a less expected location. The co-occurrence of a canonical view of horses and of light from below in the first picture makes shadow regions salient and object-like, to the detriment of objects (the horses.) In the upside-down picture, light comes from above, and shadows are expected; on top of that, shadows do not present a properly oriented view of the horses.

Although perception is occasionally mistaken about what is a shadow and what is an object, we can usually detect these errors and, when that happens, eventually perceive the shadows correctly. In this instance, perception is affected by cognition; our knowledge that the dark shapes are shadows of the horses helps perception link them appropriately despite the initial miscategorization. In the case of augmented shadows (fig. 7.12), other cues posed within the shadow shape trigger a recognition of the shadow as an object, but this alternate object seems to live a dual life with our perception of the shadow as a shadow.

Figure 7.12
An augmented shadow. Two candles activate face recognition and overcome the fact
that the silhouette of the face is not recognition optimal. Image credit: RC, August
2013.

7.6.1   Shadow-Object Chimeras

Chimeras are hybrid, cross-categorical objects. A shadow chimera is formed
each time an object gets extended in visual space through its shadow (fig.
7.13). The resulting visual item is half object, half shadow; however, it
is perceived, in favorable circumstances, as an extended object with no
shadow rather than an extended shadow with no object casting it. We may
know that this is an illusory object, but knowing does not influence the
compelling percept.

The shadow part of a chimera is perceived as a real part of a concrete object.

The same scene from different viewpoints. Resorting to less unusual viewpoints attenuates or destroys the chimera. Image credit: RC.

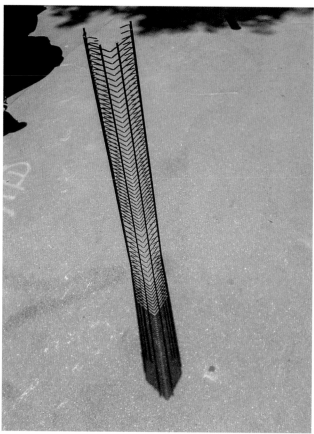

Figure 7.13
Where does the object begin, where does the shadow? Image credit: RC, Paris, June 2015.

Chimeras are visual accidents, depending as they are on viewpoint-induced visual alignments. As such, they are unstable and vulnerable to changes in viewpoint (see image in the margin). Plausibility may explain the (limited) illusory character of chimeras: it would be an unlikely coincidence if an object, or an object's part, was perfectly aligned with its shadow. If this happens, the shadow character is suspended, and the shadow merges with the object region in a chimera.

In figure 7.14, haze makes the shadow body visible within the courtyard. But the coincidental alignment of the shadow body's upper boundary with

Figure 7.14
Jardin des Plantes, Paris. Image credit: RC, May 2011.

Figure 7.15
A chimeric ladder—half object, half shadow. Image credit: RC, August 2016.

the shadow terminator on the ground and the roof's line creates a chimeric entity that takes up a volume of its own.

If accidental alignments create the strongest chimeras, nonfigural properties such as object knowledge can also play a role. The chimeric stepladder in figure 7.15 capitalizes on our knowledge of ladder shapes influencing our perception.

7.6.2 What Happens to Shadows When Shadow Character Is Lost?

If shadow labeling and character can be destroyed, what happens to the visual zones corresponding to physical shadows? They do not look like shadows anymore, but what do they look like? Shadow zones often turn into patches, that is, stable surface properties. They can turn into objects. They can even turn into holes or cracks. Much depends on what visual cues are available in the context.

Segal introduced the term *crackdow* to label a visually ambiguous pattern that could be either a cast shadow or a crack. In figure 7.16, the shadow of

Figure 7.16
A shadow turns into a crack. Image credit: Unknown artist, poster for the Studio
Arena Theatre, late 1990s.

the man ends at the border of the cliff, and the shadow's shape merges with
that of the cliff. Right next to the man's foot, the dark region may appear to
be a cast shadow, but it quickly transforms into a crack—another example
of nonmonolithic shadow character. In the region interpreted as a crack,
the reduced illumination is seen as an attached shadow on the side of the
crack. Shadow character is thus not actually lost but changed.

Failed shadows can turn into color patches, objects, cracks, or holes.

7.7 Shadows Can Create *Real* Objects

In ecological situations, the absence of sun in a shaded region can affect the ground material on which it falls. In figure 7.17, the sun falling on the playground has already melted the frost outside the region shaded by the play horse. The residual frost within the shadow destroys its shadow-ness, as the area is too light for a proper shadow. Moreover, the shadow has continued to move, so that the unfrosted leading edge and the still-frosted trailing edge add an interesting configuration, suggesting the presence of a hole in the ground.

Shadows can leave a physical impression of their presence.

In figure 7.18, a small, crescent-shaped ice strip has been created by the shadow of a fallen branch. All the ice around the shadow has melted, and only the little strip has survived. One can actually lift the strip and store it in one's fridge. Ambient conditions can bestow longer-than-usual lives to the association of shadows and permanent surface features.

Figure 7.17
Hole shadow. The sun outside the shadow melts the frost, leaving a light-colored frost shadow within. The real shadow keeps moving from the bottom to the top of the image, and the slightly offset combination of real and frost shadow appears more as a hole than a shadow. Image credit: RC, Monza, Italy, January 2008.

Figure 7.18
Can shadows create objects? Image credit: RC, Etna, New Hampshire, 2014.

7.8 Analogues of Shadows

The representations of shadows passed on to conscious cognition are primarily, if not exclusively, visual. We may associate temperature feelings with shadows (we are in the sun, and the shadowy spot under an oak tree affords freshness), but there do not seem to be auditory or olfactory sides of shadow representation. However, other senses create analogies to visual shadows, for example, "auditory shadows." The treatment of shadow information by cognition provides for metaphors and figurative language, and these too deserve exploration. The notion of a shadow can be extended to cover cases in which a configuration results from interrupting a flow—not only a photon shower but any kind of flow, such as snow, rain, urine, or stardust. Thus an umbrella can create a shadow in rain—where you can get shelter. The cast rain shadow of the umbrella is normally lighter than the surrounding wet area. A table can create a snow shadow. The cast snow shadow of the table is in the norm darker than its surroundings and may look like an ordinary

Figure 7.19

In some conditions, the analogical shadow may be mistaken for the standard shadow. Here we see a standard shadow in light, and an analogical shadow in snow, both created by the same object. Image credit: RC, Etna, New Hampshire, February 2014.

shadow. Sometimes you can have both a standard and an analogical shadow in the same scene, as in figure 7.19 and in the margin. Technical extensions of the notion of shadow are documented in glaciology and in acoustics. However, these and more distant metaphors are beyond the scope of this work.

The notion of a shadow covers other gaps in flows.

Two types of shadow a pony may cast. Image credit: RC, Paris, 2016.

7.9 Conclusions

In chapter 6, we presented visual conditions for the attribution of shadow labeling. In the present chapter, we explored further how the perceptual and cognitive systems reason about shadows. There are a number of studies on the boundary between the cognitive system and the perceptual systems and other systems (such as the motor system). What we know about

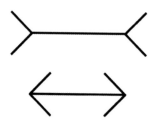

You can measure, and hence know, that the horizontal bars are of equal length, but this does not interfere with your perceiving them to be of unequal lengths.

the mechanisms of the cognitive system is far from being as detailed and well-framed theoretically as our knowledge about vision. Quite a few examples demonstrate the independence of cognitive and visual knowledge. In the case of shadow perception, we have anecdotal examples of "cognitive impenetrability," where a cognitive decision (e.g., that a region is not a shadow) does not affect the perceptual decision (the use of the shadow information from the region). The perception of dark pigment as shadows in photographs and paintings is an obvious example where our knowledge that these are not actual shadows does not impede the shadow inferences that we extract from them. However, we also showed instances where the knowledge that a dark region is a shadow could correct its initial perception as an object. This influence of knowledge on perception is probably limited to cases where the percept is already weak or bistable.

Overall, cognition has a broader conceptual framework for shadows than does perception, and though we can usually detect errors made by perception, cognition itself does have some ill-conceived notions about shadows. As usual, the big advantage of cognition is that it can learn new concepts and concept boundaries, whereas shadow rules and concepts for perception appear to be fixed, perhaps even from birth (e.g., the bias for light from above seen in chickens raised from birth with light from below).

We have seen that the properties used for making inferences about shadows—causal, figural, visual—interact in different ways in different contexts. In some visual conditions, a shadow can appear as having been cast by another shadow (fig. 7.3). In the situation we described, the shadow on an opaque screen was looking for an object to be a shadow of, and the only available object was the shadow cast on a semiopaque screen. Generalizability of these effects out of their original context appears difficult. Here is the patchwork of local rules based on the examples we discussed.

• Sometimes perception is in error, and cognition detects the problem.

Whatever looks like an opaque object is expected to block light and to cast shadows. Even an illusory object generated by a shadow is expected to cast a shadow.

• Cognition has a broader concept of shadows than does perception.

For perception, a shadow must have a visible boundary.

We can reason about shadow bodies, but unless they are made visible by dust or fog, they are not used by perception.

• For some situations, cognition is ambivalent about what is or is not a shadow even though perception is not.

A filter shadow acts as a shadow for perception but is ambiguous for cognition. Under an extended light source, the shadow may lose its resemblance to the object that casts it.

• In some situations, cognition initially has a misconception about shadows where perception does not.

Cognition may treat shadows as things that can be projected or blocked.

• Perception favors object interpretations over shadows, where possible.

When a shadow projects a more canonical view than the object itself, the shadow may be perceived as an object.

The shadow part of a chimera is perceived as a real part of a concrete object.

Failed shadows can turn into color patches, objects, cracks, or holes.

Shadows can leave a physical impression of their presence.

The notion of a shadow covers other gaps in flows.

In this chapter, we also examined the terminology used to describe these phenomena, hinting at the limits of commonsense lexicalization, cross-cultural variation (absence of the shadow/shade distinction in some languages), the range of application of technical language ("umbra"), and the necessity of introducing new descriptive terms ("filter shadow").

Further Research Questions

1. Does the shadow-pigment ambiguity call on mechanisms specific to shadows? Do other cases (light pigment) share computational mechanisms?

2. How far can shadow metaphors stretch? A pedestrian can hide in the shadow that a parked car creates in traffic—behind the car, relative to the traffic. But a traffic-free zone appears in front of the car, too: could this be a "forward" traffic shadow?

3. Shadow bodies at sunset appear to radiate in all directions from the sun, as if the sun were sending rays to a direction normal to the ground. But actually (see chap. 1) sun rays are parallel to the ground at sunset. This is a cognitive sun illusion arising from our inability to understand perspective. The sun's rays should be seen as a flat ceiling above us.

4. What is our mental model of a rainbow? What would subjects draw, if asked to trace actual sun rays in a scene containing a rainbow? The correct answer would look surprising to some.

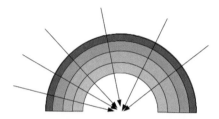

Figure 7.20

Top: The sun rays that generate the rainbow converge to the antisolar point. *Bottom*: Only a small portion of the rainbow is visible, along with shadow bodies pointing to the antisolar point. Image credit: RC, Paris, May 19, 2015.

Objects under sunlit canopies present interesting pictorial challenges. Nomellini is one of the few artists who investigated patterns of dappled sunlight in his work. Plinio Nomellini (1866–1943), *Il figlio* (detail), 1907. Novara, Galleria d'Arte Moderna Paolo e Adele Giannoni, Donazione Giannoni. Image credit: Fotografia di Giacomo Gallarate/Centro di Documentazione Musei Civici di Novara.

8 Illumination

Shadow and light celebrate an unbreakable marriage. So it will not be inappropriate, in a book consecrated to shadow perception, to say a few words about some aspects of the perception of light. Indeed, we can ask many of the same questions about light and illuminated regions that we have asked about shadows. Is that area light or paint? Does it have "light character"? Who owns the light spot on a surface? Does it provide information about spatial layout?

Between each light spot and its darker surround or each shadow and its illuminated surround is the illumination border that they share (fig. 8.1). The properties of the border determine if we take the luminance change to be a change in illumination or a change in surface material. The properties we outlined for acceptable shadow borders hold as well for the border around a light spot, as they are really properties about the same border. To rephrase the properties, then, the light spot should be lighter than its surround along its border, it should not be opaque (it must have X-junctions, not T-junctions with surface contours), it should not appear to have any volume of its own, and its border should not align with other boundaries or contours in the scene.

Violating some of these properties can induce the disruption of light spot labeling and character. If a light spot can be linked to its source, this ownership relation, like that for shadows, can support inferences about scene layout. And as for shadows, there are some conceptual intricacies. Cunning manipulation of essential properties can induce light spot labeling when no light spot is present. The chapter closes with some forensics of light and a brief discussion of how light rays are used in computer graphics to create photorealistic renderings of scenes.

Figure 8.1
Left: Patches of light on the wall appear to belong to the adjacent lamps. The light spot is smaller than the enclosing, less-illuminated region, so we don't consider the darker regions as shadow. Instead we take the light patch as the scene feature. *Right*: As the object owns its shadow, so the light source owns its spot. Image credits: Left: PC, Paris, March 2017. Right: RC, Paris, January 2018.

8.1 Light Sources

8.1.1 Light Shapes

Illuminated areas are created either by the shape of the original light source or by occluding surfaces that block the light source (fig. 8.3). If the illuminated area has the necessary properties (lighter, flat, not opaque, X-junctions), then it is labeled an illuminant effect, a light spot, and the material properties of the surface are corrected for the added light.

Projections of light through windows, or reflections from water or mirrors, also add an aperture to the light shape. In figure 8.4, we see these projections from windows onto the adjacent ground. It is the sun that gets reflected in the windows to produce the light spots. In some cases, as here, these projections are seen as images of the objects (the windows) from which they are projected, but in other cases they are not. In figure 8.4, and earlier in figure 8.2, both the overall shape of the window and some details of its internal structure are made available in the light spot. Some languages have a name for reflections that are seen as copies of their aperture (the window) but not of their source (the sun); Italian, for instance,

Figure 8.2
Light spots. Image credit: PC, Roques, September 2014.

Figure 8.3
Light is often shaped by apertures between the source and the receiving surface. Once the region is identified as a light spot or area, its illumination can be discounted. Thus, at the bottom, the two squares *A* and *B* send equal amounts of light to the observer; but *B* is seen as a dark square, and *A* as a light square, once the illumination is taken into account. Image credit: PC.

Figure 8.4
Projections of light reflected from windows of the Orangerie in Paris onto tree shadows. Image credit: PC.

has *gibigianna* and *colpo di luce* (light shot), typically projected on ceilings by wristwatches under the sun.

Light spot shape mimics the aperture shape.

8.1.2 Mutual Reflection

Highly reflective surfaces produce well-defined light spots, but every surface reflects some light onto neighboring objects (fig. 8.5). This creates mutual reflections or interreflections that can sometimes be noticed if they have a different color. Because they are diffuse reflections, their intensity

Figure 8.5
Mutual reflections or interreflections originate from the diffuse reflection of a light from one surface to a nearby surface. Here the reflecting surface is yellow, slightly coloring the adjacent skin. English folklore has it that if this particular flower, the creeping buttercup (*Ranunculus repens*), casts a yellow reflection on your skin, then you must like butter. Image credit: Vignolini et al. (2011).

drops with distance from the reflecting surface, giving a cue to the distance between surfaces.

In most scenes, especially indoors, much of the light comes from reflections of light off the surfaces in the scene. These mutual reflections are strongest when two surfaces are close to each other, as in the case of the flower held under the chin, but also where surfaces meet, in the corners of a room, or where an object sits on the floor. Light bounces back and forth between these nearby surfaces, creating a gradient of diffuse illumination. We often do not become aware of these interreflections other than in extreme cases. Nevertheless, the visual system has some knowledge of their presence and influence. To demonstrate this phenomenon, Gilchrist and Jacobsen built two identical rooms, one where everything was painted white, and the other where everything was painted black (fig. 8.6). They then adjusted the lighting to each room (the light sources were not visible) so that the average luminance coming from each room was the same. Observers nevertheless easily reported that one room was white and the other black. This result shows that the pattern of mutual reflections had filled the surfaces in the white room with a diffuse light, reducing the overall contrast. In the black room, there was little mutual reflection, so the

Figure 8.6
A smaller version of Gilchrist and Jacobsen's white-room-versus-black-room demonstration. Everything on the left is painted white, everything on the right is painted black, and the lighting is adjusted to give the same amount of light returning from both. The surfaces on the left appear to be painted white, whereas those on the right appear painted a darker shade. (The effect is much stronger in the real versions than in the photos.) Note the differences in the depths of the shadows between the two. This is due to the interreflections being much stronger off the white paint and is probably the cue to the visual system for the surface reflectance. Image credit: Alan Gilchrist.

directly illuminated surfaces had extremely high contrast with the regions in shadow. Observers did not have to reason about these mutual reflections to judge the paint color in the rooms; they simply saw white or black surfaces, indicating that the visual system knows about the mutual reflection effects at the perceptual level.

Perception uses mutual reflections to recover the reflectance of surfaces.

8.1.3 Flat Light Spots versus Light Bodies

As we can with shadows and shadow bodies, so we can consider the light spot as an isolated two-dimensional projection of the source or as one end of a light body that joins the source to the light spot and is bounded along its volume by regions where the light has been blocked. When atmospheric conditions allow, it is possible to see light bodies that are also referred to as *light ray traces*. Light rays in the norm are not visible, except perhaps for the case in which one stares at a light source, unless intervening particles—for example, water or dust—scatter the light along its path (fig. 8.7). At least one language, Québécois, lexicalizes the phenomenon (using the expression *pied-de-vent*, literally "foot of the wind," when the rays reach the ground). Light bodies and shadow bodies must alternate when they are visible, and

Figure 8.7
An artistic representation of atmospheric light bodies. Giovanni Battista Tiepolo, *Apollo and the Continents* (detail), 1752–1753. Fresco, stairwell of the Residenz, Würzburg. Image credit: Web Gallery of Art.

whether light or shadow bodies (or both) are singled out depends on the relative volumes involved (fig. 8.7).

As we mentioned in chapter 1, a visually compelling and seldom remarked phenomenon is the convergence of the sun's light ray traces at the antisolar point. As the rays are physically parallel (at our distance from the sun) and at sunset and sunrise are parallel to the ground, if we turn our back to the sun, we can see them visually converge in a point that is opposite to the sun.

Light bodies reveal the direction of the illuminant.

Figure 8.8
The standard icons of the sun capitalize on the visibility of light bodies and shadow bodies, as this simple web search for "sun icons" shows.

8.2 Light Ownership

8.2.1 Linking Light Source and Light Spot

Computations about the ownership of the light source casting the light spot are similar to those for the ownership of the object casting a shadow, although, necessarily, light sources are more distant than the objects that cast shadows. If a visible light source illuminates a small region, it may appear to us that the light owns the bright patch. Proximity may be the critical property (fig. 8.9), although other cues, if present, to the trace of the light beam (light bodies) may overrule proximity. Light sources typically do not have shapes that are copied in their projected light spots, so the copycat principle that was effective for shadows is seldom a factor. However, projected light shapes occur when a light spot is created by a hole in an occluding surface with light shining through it, like a doorframe or window. In this case, we may see illuminated shapes that match the aperture, patches of light that look like the windows that let in the light, for example (see again figs. 8.2, 8.4).

Light spot ownership is most often determined by proximity to an aperture or source.

Figure 8.9
The light spot on the wall comes from the nearby window. Image credit: Eugenijus Radlinskas.

8.2.2 Misattribution of Light Ownership

Perception (and cognition, as we saw in chap. 7) may be uncertain about some properties of shadow. The same holds for light, and we can see this in relation to the problem of light ownership. When two or more light sources exist, the color of the shadow can differ from that of its surround. Cast shadows on snow are blue, as they receive ambient light from the blue sky and none from the yellow sun; the surround is white, receiving both the yellow sunlight and blue sky light. But in the presence of multiple light sources, the attribution of the color in the shadow can be puzzling. This defines a *light ownership* problem. Consider figure 8.10, which displays red shadows and green shadows cast by a fence. The star-shaped symbols indicate the position of two possible light sources. Which of the two is the red light source, and which is the green light source?

When asked this question, subjects typically place the red light source on the top, and the green light source on the bottom. It is as if a red source

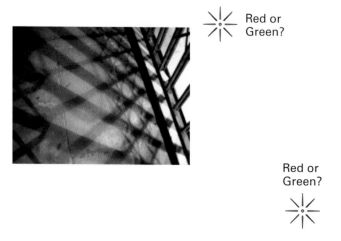

Red or
Green?

Red or
Green?

Figure 8.10
Where is the red light? Where is the green light? Image credit: RC, Florence, Italy, 2006.

would naturally be associated with the red shadows that splay out from the direction of the source, as if colored shadows are interpreted as light beams. However, the physical situation is the other way around. Red light gets to every place in the figure except the areas where it is blocked by the bars of the fence. Those areas receive only green light. Thus the green shadows are actually produced by the red light, at the bottom of the figure.

The counterintuitive association between red light and green shadows, and vice versa, hints at a strong mental link between shadows and light sources, as if the light owns the shadow of its own color. Of course, the casting object owns the shadow, and the light owns the illuminated areas, but this logic is not that of our cognitive system.

Perception does not know what color a shadow should be.

Nor does cognition, for that matter.

8.2.3 Light Spot Disruption

If an area qualifies as a light spot because of its luminance, it is not thereby automatically seen as a light spot. Absence of X-junctions and manipulations of the boundary of the light can have disruptive effects similar to the ones we observed for candidates to shadow labeling or character.

Light spots cannot have an outline.

8.2.4 Conceptual Intricacies: Crazy Lamps and Holes in Shadows

In figure 8.12, the lamp appears to own the light falling on the nearby wall. However, the lamp also owns its shadow on the wall. Clearly something is amiss. Some lamps do cast shadows of their frames. (We saw an example of this in chap. 5, when discussing a streetlamp depicted by Magritte.) However, no lamp can cast a shadow of its entire shape, and yet we are drawn toward this erroneous interpretation. The problem arises because of the presence of two light sources, a situation that challenges the cognitive system, as we saw in the Gelb demonstrations of chapter 3. The lamp does own the shadow, but the lamp does not own the light that casts the shadow.

Proximity can lead to erroneous light ownership.

Shadows are holes in light, but what about a hole in a shadow? It creates a light spot, so that a hole in a hole in light turns out to just be a spot of light (fig. 8.13). The concept of a shadow appears to be used for talking about projections: we do not find it inappropriate to describe the hole in the shadow as a shadow of the hole in the frame.

Figure 8.11
As with shadow disruption, so illumination disruption threatens the perception of light spots. With no X-junctions and a highlighted dark boundary, the light from the lamp appears to be a permanent color area. It becomes "wooden," in Leonardo's terms. Image credit: RC, Vieques, Puerto Rico, February 2014.

Figure 8.12
Crazy lamp. Image credit: RC, Etna, New Hampshire, March 2014.

Figure 8.13
A hole in a shadow creates a light spot. Image credit: RC.

8.3   Scene Layout from Light Spots

When a light spot appears to have an owner, it releases information about the location of the receiving surface relative to the source. The adjacency property for light spots and their sources is equivalent to that for cast shadows that Leonardo noted. If the light spot contacts the source, then the receiving surface is also in contact with the source. In contrast, if there is a gap (fig. 8.14), then the receiving surface is not in contact with the source.

If a light spot touches its source, then the surface on which the light falls also contacts the source.

While these light-based inferences about layout may be perceptual, others are available that are purely cognitive. For example, tree canopies often create pinholes between groups of overlapping leaves. As a result, within the overall dappled shadow of the canopy, there are small spots of light that

If the sun is not directly overhead, or if the ground is slanted, then the images of the sun will be elliptical; but if X is the width of the shorter axis of the ellipse, then the height of the canopy is approximated by the following expression (where the subtended arc is the product of the radius and the angle in radians):

$$X/(0.5 \bullet 2\pi/360)$$

So a light spot that is 20 cm across its smallest width implies a canopy at about 23 m. This piece of information could be used in aerial photography to estimate the height of canopies over ground.

Figure 8.14
Left: The window owns the light spot, and the gap between the window and the light spot indicates that the window does not sit flush with the floor. *Right*: In contrast, the light from the door and the light falling through the door onto the floor meet, indicating that the floor is flush with the door—there is no sill or step. Image credits: Left: RC.

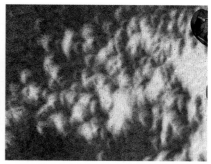

Figure 8.15
Three pinhole-generated images of the sun on the ground. *Left, middle*: The larger the image, the more distant the pinhole and the canopy. *Right*: Pinholes project an image of the sun, as is evident from their appearance during a partial eclipse of the sun. Image credits: Left, middle: RC, Hanover, September 2013. Right: Josée Rivest, Toronto, August 2017.

are images of the sun on the ground (fig. 8.15). The lensing properties of pinholes were described in the sixth century BCE in China and then again in the tenth century CE by Alhazen. The lens properties are more obvious during a partial eclipse when the images cast by pinholes in the canopy become crescents (fig. 8.15, right). Since the visual angle subtended by the sun is a constant 0.5°, the size of the image falling on the ground below the canopy is a function of the distance between the pinhole and the ground. Larger images of the sun imply a higher canopy.

In the dappled light, the larger the individual spots of light (each is the sun's image), the higher up the canopy.

Many painters were fascinated by the challenge of representing objects under canopies, but few paid attention to the exact properties of pinhole-generated light spots. One example where the painter did properly capture the dappled light effects is the Tuscan "divisionism" painting in the picture at the beginning of this chapter.

8.4 Faux Light Spots, Not from Light Sources

When some light material like snow accumulates through an aperture, it may appear as a light spot. Equivalently, when an accumulation of darker material is blocked by an intervening object, it also creates a light area that appears to be a "light shadow." Given the nature of light and opaque objects, there are no real light shadows, of course. Full-fledged faux light shadows

Figure 8.16
The elliptical sections of the cones generated by pinholes have all the same width along their shorter axis in the scene (although, depending on perspective, not necessarily in the image), independent of the slant of the surface on which they are projected. Image credit: RC, San Francisco, May 2014.

are ecologically possible (fig. 8.17). It also turns out that in visual search tasks, light shadows are not discounted as normal shadows are, consistent with the darkness constraint for shadow character discussed in chapter 4.

Faux light spots occur when a dark flow is blocked.

8.5   Forensics of Light

We saw in chapter 4 that human observers are not good at detecting inconsistent illumination. This critical weakness makes us excellent targets for the flood of manipulated images that bombards us on the Internet. Many images in advertising, news photography, Internet hoaxes, and of course propaganda are artfully or sometimes clumsily altered. Although we often cannot detect the fakery, the information is there. Computer analysis can distinguish unaltered images from doctored ones specifically because the

Figure 8.17
Faux light spot in the area under the bike shielded from rain. Since water acts as a filter, it darkens the surface and creates a contrast with the shielded area. Image credit: RC, Paris, July 2016.

effect of light and how it falls on objects is so constrained. Here we present two examples where light analysis reveals the addition of different people who were not in the original photograph. In the first (fig. 8.18), the highlights on the eyes do not match as they must across the individuals. Clearly, some people were photographed under different lights. In the second (fig. 8.19), the distribution of light over the four people is analyzed to recover the light source direction. The direction for the man on the left does not match. These examples show that each scene contains an enormous amount of lighting information, but as we have noted, little of it is used by the human visual system to understand scene layout, undoubtedly because of the enormous computational burden it would place on the brain. The few properties that our visual systems do analyze have the advantage that they can be analyzed rapidly and are typically sufficient. Our new environment of photography, art, and the Internet can slip many inconsistencies unnoticed past our simpler scene-understanding analyses. This has advantages for the liberties that artists can take, but also costs for the visual fraud that escapes our notice.

Computers can check for lighting consistencies far beyond those checked by the human perceptual system.

Figure 8.18

The judges of a popular TV program sit for a group photograph in the theater. It looks like an ordinary photo, but close-ups of their eyes reveal at least two different highlight patterns. If they had all sat for the same photograph, the same light would be illuminating all of them, producing similar reflections in their eyes. Image credit: Johnson & Farid, 2007a.

Figure 8.19

The three gentlemen on the left appear to be photographed in the same light, and the person in the middle even casts a shadow on the person on the left. However, an analysis of lighting direction shows that the man on the left has a different illuminant direction from the other two. Image credit: Johnson & Farid, 2007b.

8.6 Ray Tracing in Computer Graphics

Computers can analyze the light patterns in actual scenes, but they are also able to generate scenes that are strikingly photorealistic. The accuracy and realism of computer-generated graphics have grown over the past decades until it is now commonplace for most movies to have significant portions generated by computer, which, except for the obvious magical aspects, are completely perceptually convincing. Some challenges remain in the treatment of human skin and issues of kinematics, but at least for light and shadow, little is left to be added in terms of realism. Compared to the 1/10 of a second processing time for human scene understanding, many of these computer rendering efforts may take hours or days to complete each frame of a sequence. Many of the current advances focus on rendering complex scenes in real time by taking shortcuts in the modeling of light and reflection. The goal of these approaches is to lose realism where human observers will not notice, thus taking advantage of shortcuts in human visual cognition. For example, one study found that rendering simplified to only 1 percent of its original complexity generated cast soft shadows that satisfied 90 percent of observers.

A complete rendering of the light in a scene would start with each light source in the scene and trace all their light rays going in all directions, reflecting from surfaces, diffusing through translucent materials, refracting through transparent media. However, the vast majority of these rays never reach the eye. So the *ray tracing solution* creates the image by starting from the eye and tracing rays backward to the sources, determining what light will be arriving on each ray and placing that value on the image location that the ray intersects (fig. 8.20). If a ray strikes an object, then a second "feeler" ray is traced from that location to the light source to determine if the point can "see" the light source. If that feeler ray to the light source is unobstructed, then that point is illuminated, and the brightness and color are computed. If the feeler ray is obstructed, then that location is in shadow, at least for that light source. This approach has many variations, and one, called radiosity, addresses diffuse light reflection from surfaces rather than individual rays of light.

Computer-generated images may model light quite accurately but often take shortcuts chosen to be less noticeable by human vision.

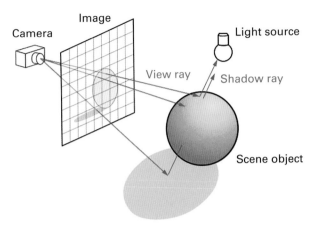

Figure 8.20
Rendering scenes with ray tracing. Rays are traced backward from the camera to the light source. If a ray intersects with an object, a second ray then tests whether that location sees the light. Image credit: Henrik, Wikimedia Commons.

8.7 Conclusions

In this chapter, we explored how light can form entities, light spots, that act like shadows in many ways. These follow much the same logic as shadows, except for the inversion of light polarity. Thus light spots are areas of light surrounded by shade, which trigger corrections for the illumination, may appear to belong to specific light sources or apertures, and have some influence on understanding scene layout. We also hinted at some of the indeterminacies of the concept of light and of light spot. Finally, we covered some of the computer vision and computer graphics approaches to analyzing and rendering light.

Light spot shape mimics the aperture shape.
Perception uses mutual reflections to recover the reflectance of surfaces.
Light bodies reveal the direction of the illuminant.
Light spot ownership is most often determined by proximity to an aperture or source.
Perception does not know what color a shadow should be.
Light spots cannot have an outline.
Proximity can lead to erroneous light ownership.

If a light spot touches its source, then the surface on which the light falls also contacts the source.

In the dappled light, the larger the individual spots of light (each is the sun's image), the higher up the canopy.

Faux light spots occur when a dark flow is blocked.

Computers can check for lighting consistencies far beyond those checked by the human perceptual system.

Computer-generated images may model light quite accurately but often take short-cuts chosen to be less noticeable by human vision.

Further Research Questions

1. Are light patterns ignored or demoted to the same extent as shadow patterns? Would visual search for an odd direction of light (say a set of small vertical mirrors all making reflections in front of them) be as difficult as it is for an odd direction of shadow?

2. Forensics of light: Can human observers be trained to detect lighting anomalies? How good are professional lighting specialists?

Being a mirror and looking like a mirror. The impression that a mirror is present in the scene is not disrupted by a number of incongruities between the scene and its reflection. The image in the mirror is contrast reversed and the frown changed. The mirror still looks like a mirror despite the mismatches. Image credit: PC, Paris, September 2017.

9 Shadows, Reflections, and Transparency

9.1 Introduction

This chapter deals with two interrelated topics. There are interesting similarities and differences between the information in, and the visual processing of, shadows and that of reflections and transparency.

Both shadows and reflections preserve some of the visual properties of the *source object*, that is, the object that casts the shadow or that gets reflected. In particular, an object's cast shadow is a projection of its silhouette as seen from the direction of the light source, whereas an object's reflection gives all the visual properties of the object as seen through the reflecting surface. The visual system can use both shadows and reflections to recover properties and the location of the source object and access information from viewpoints other than that of the viewer. In both cases, the visual system relies on a simplified logic to retrieve and use the relevant information. Both shadows and reflections are sometimes missed by vision. Finally, the disruption of shadows by outlines or misalignment, for example, has parallels in some similar disruptions of "mirrorness."

Shadows and transparent surfaces share yet another set of properties. Both depend on the availability of X-junctions, and both split an area in the scene into shadow and nonshadow regions, or regions seen or not seen through a filter. In each case, the reduced luminance returning from the shadowed or filtered region is attributed not to a darker surface material but to the reduction of the light blocked by an opaque object (causing a shadow) or filtered out by a less than perfectly transparent layer. Finally, shadow disruption has some parallels with transparency disruption.

Given these similarities, a general question is therefore whether shadow perception is an independent processing unit or whether it shares computational resources with reflection perception or transparency perception.

A couple of caveats. In this chapter, we use the term "reflections" to refer to the visual content revealed by a mirror. In contrast, the technical literature often refers to "mirror *images*," but this falls into conceptual and terminological controversies concerning pictures and images and uncertainties about where we see the reflected objects and space. We also restrict our discussion to flat mirrors. Reflections from concave and convex mirrors are fascinating and pose interesting perceptual problems of their own, but their informational structure has few equivalents in shadow casting. Similarly, we restrict our discussion principally to flat transparent materials lying directly on opaque surfaces below them. Our section on filter shadows (chap. 7) already covered the shadows that filters cast when they are not flush with the surface behind them.

In covering both reflections and transparency, we follow the general approach we used for shadows and illumination: first, what is the information available; and second, how is it used to understand the scene (the mission)? Then we ask if there is a process to establish ownership: what object is paired with what reflection? Then, what is it to be a reflection, or a mirror, or a transparent overlay? Finally, what are the properties required by the human visual system to label a surface a mirror or a transparency, and are there physical properties that can be ignored, or conversely properties that act disruptively?

9.2 Reflections

Mirrors are informational tools. They reflect light at an angle to the surface that is the same as the incident angle. This creates a visual scene within the mirror frame that matches what is in front. They also, importantly, provide a different viewpoint on the scene: someone looking at a mirror sees what a symmetrically placed observer would see through the mirror from the other side if the mirror were transparent (modulo a left–right swap).

What information do mirrors provide? We quickly list some here, to match our treatment of the information in shadows (chaps. 1 and 2). Unlike the reduced information provided by shadows, objects in mirrors are as fully specified as directly viewed objects in terms of color, three-dimensionality, and motion parallax. Nevertheless, observers face challenges in retrieving the location and orientation of the actual objects seen in mirrors that are similar to the difficulties faced using a shadow to recover the location and orientation of the object casting a shadow.

Mirrors make a wealth of extra information available to the observer from the additional viewpoint, like the possibility of simultaneously seeing both front and back of an object, or seeing behind our back (in detective movies, we see the villain approaching in the reflection of someone's sunglasses in front of us), or seeing our face in the mirror or our own profile with two mirrors, something we could not have seen otherwise and could not have predicted either (other than by photography or painting). Some learning is necessary to use this information, and furthermore some of this is made available only through thinking. It is thus interesting to sort out what information we can extract from mirrors without thinking—the way shadows tell us about spacing and contact without us having to think about it. Here the mirror provides more information than a shadow, which conveys only the object's silhouette from the viewpoint of the illumination.

Do mirrors exist in natural settings? Calm water surfaces provide plenty of natural mirrors (fig. 9.1), and one may expect that some perceptual sensitivity to mirror reflections coevolved with the visual system, for example, discounting the mirror image of a bird on water (only *one* bird is worth our cognitive investment; recall how shadows are demoted). Animals attack their vertical reflection but not their reflection on mirrors on their cage floor, and there are countless anecdotal reports of birds that die flying into the sky seen reflected in mirrored buildings, but no reports of birds flying into the sky seen reflected in still water. The ecology of natural mirrors (still water) suggests that they must have been noticed in the course of the evolution of vision, and some discounting rules should be part of our biological endowment; conversely, as reflecting water surfaces lie on the horizontal plane only, we need to ask what we understand of mirror reflections caused by the cultural, nonnatural vertical mirrors we mostly use today.

As mentioned, some learning is necessary to use the information provided by mirrors. Cunningly placed mirrors in homes and restaurants can generate a perfect and inescapable trompe l'oeil illusion of space, but most mirrors do not, and they are seen and used for what they are: cognitive devices. On top of learning to attribute the reflection to its source and to understand its orientation, we have to learn that all the space that appears

Figure 9.1
Discounting one of two birds. Image credit: Rich Cavanagh.

to be on the other side of the mirror is actually on our side—so that we should not walk into the wall! We have to learn how to steer our movements as we look at our reflected limbs. As drivers of vehicles, we learn how to make quick decisions related to things that happen behind us based on what we see in rearview mirrors placed in front of us. Notably, the sense that the objects in the rearview mirror are behind us is learned as well: objects initially appear to be off to the side, as is reported when driving for the first time on the other side of the road with the rearview mirror on the other side as well (e.g., continental Europeans driving in the United Kingdom). Most of this is acquired through sensorimotor learning. In the case of shadows, we have seen that the offset between an object and a shadow does help to locate the object relative to the surface on which the shadow lands. However, there seems to be no shadow equivalent of the rearview mirror, a device built to use shadows for locating objects that cast them (but see a fanciful version in the margin).

We can draw many analogies between shadow perception and mirror image perception. On the information side, both shadows and mirror reflections are a kind of projection, and both provide the viewer with extra information about an object. On the vision side, in the case of shadows, we have already described (chaps. 1, 2, and 5) how viewers are sensitive to some shape/shadow correspondences while being systematically blind to others. A similar tolerance is found for impossible reflections, along with the rejection of accurate reflections for some mirror contexts but not others. Part of the reasoning involving mirrors includes the properties that determine what will be seen as a mirror by a human observer: properties of "mirrorness." This set of properties has similarities to that for determining what is seen as a shadow. In the first half of the chapter, we consider these similarities in the treatment of shadows and mirrors, and whether this similarity suggests that shadow perception and mirror perception use common computational resources.

As mentioned, we will limit the discussion here to plane mirrors. A few more warnings are in order. We will also keep in mind that the perception of real mirrors is not the same as perception of mirrors depicted in artworks. Depictions of mirrors strip down the stimulus by copying only some of the informational aspects of real mirrors (no motion parallax, 3-D, no real scene in the mirror) and reveal the visual system's rules for deciding what is a mirror and what is an acceptable reflection in that mirror.

What we see in a mirror is viewpoint dependent, as if there is a copy of the room around us, flipped back to front and visible through a mirror "window." As we move around, the mirror reveals different parts of the virtual room through that window. The virtual objects in the mirror are flipped back to front around the mirror's surface as a symmetry plane, so that point A of the real object is at the same distance from the mirror surface as its virtual counterpart A', along the line normal to the surface that connects the two points. The virtual object has no physical reality at the location where it is seen; it is "as if" we are seeing it.

We can also consider the intersection of the mirror surface and the light as it travels from the object to the observer's eyes (fig. 9.2). We will call this intersection the "surface projection." We do not perceive the surface projection, and we are surprised to discover its actual size on the face of the

An (imaginary) device that does with shadows what rearview mirrors do with light. The presence of objects behind the viewer could be inferred by the shadows they cast in front. Image credit: RC.

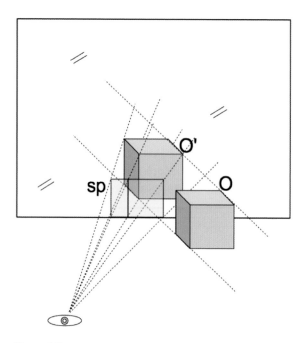

Figure 9.2
The virtual object O′ is the reflection of object O, as visible from the viewer's location. The intersection of the light from the virtual object to the observer defines a surface projection (*sp*), which is not visible. The virtual cube and the real cube are the only visible items. Image credit: RC.

mirror; this is particularly evident when the reflected object is the viewer, as Gombrich pointed out:

It is a fascinating exercise in illusionist representation to trace one's own head on the surface of the mirror and to clear the area enclosed by the outline. For only when we have done this do we realize how small the image is which gives us the illusion of seeing ourselves "face to face." To be exact, it must be exactly half the size of our head. … But however cogently this fact can be demonstrated with the help of similar triangles, the assertion is usually met with frank incredulity. And despite all geometry, I, too, would stubbornly contend that I really see my head (natural size) when I shave and that the size on the mirror surface is a phantom.

Tracing the profile of the reflection of one's own face on a mirror (e.g., when steam covers its surface) reveals that the projection is exactly half the size of the face itself. This is a consequence of the fact that the surface is always exactly halfway between the object (one's face) and the virtual object (the face's reflection).

9.2.1 Information in Reflections

9.2.1.1 Location and the "times two" rule   Reflections, just like shadows, involve a surface other than the surface of the object reflected or projected. This entails that reflections serve anchoring and position functions quite analogous to those provided by shadows (see chap. 1). For instance:

If an object does not touch its reflection, it does not touch the reflecting surface, either.

An object that touches the reflecting surface must touch its reflection. This matches the properties of contact for cast shadows (chap. 1). As in the case of cast shadows, the converse rule is not in general valid, as its validity may depend on contingencies of viewpoint. In the image, an object may touch its reflection and still be separated from the surface of the mirror.

However, let us note two crucial differences. First, consider the reflection of a ball touching a wet surface (fig. 9.3). A reflection of an object must

Figure 9.3
Reflections' anchoring function. Note that in the image the reflection is smaller than the ball. Image credit: RC, Paris, December 2014.

always be visually smaller (subtend a smaller visual angle) than the object in direct view because it is necessarily farther away, "behind" the reflecting surface. But a cast shadow can often be visually larger than the object that casts it in the image. As we saw in chapter 2, the smallest cross section of a shadow body cannot be smaller than the smallest cross section of the object casting the shadow (assuming sharp, not diffuse, light). So considering cast shadows from smallish sources (like the sun, which is 0.5 degrees in the sky):

An object's reflection is always smaller than its shadow.

Second, reflections are the same distance from the reflecting surface as the object being reflected:

A reflecting surface is halfway between the source object and its reflection.

This means that if used as position indicators, reflections invariably are at a "times two" distance from the reflected object (fig. 9.4). On the other hand, the distance between an object and its shadow depends on the direction of

Figure 9.4
Birds over water fly lower than they appear. How distant is the bird from the surface of water? If the distance from the bird to its reflection is *d*, the distance from the reflecting surface is *d/2*. Had the reflection been a shadow instead, the distance from the surface would have been *d*. Image credit: http://www.photos-public-domain. com/2012/04/27/two-geese-flying-low-over-water. License: Creative Commons CC0. Accessed September 24, 2017. Graphic modification: RC, PC.

Figure 9.5
Shadows and reflections in a single image. Because the shadows are cast on the reflecting surface, they are in between each object and its reflection that is vertically below each bird. Image credit: *James's Flamingos over Laguna Colorada, Bolivia*, photographer unknown.

the light source (fig. 9.5). Only where the light is normal to the surface (e.g., when the sun is directly overhead) is the distance to the shadow equal to the shortest distance to the surface (a times one rule).

9.2.1.2 Visible properties and viewpoint
Reflections can reveal aspects of the world that are not available from our usual viewpoint. This is the mission of mirrors. This is something that we saw for shadows too, but clearly in the case of reflections, it is a defining feature: we buy mirrors mainly for this reason.

Reflections allow us to acquire information about properties of an object that we could not otherwise see. In some instances, as described in the next section, reflections also allow us to navigate in the world from a different viewpoint (think of driving in reverse using the rearview mirror). The two functions are intertwined but differ in how we use the mirrored scene, passively for inspection or actively to navigate.

In the first case, we see properties of an object that are on its far side, not visible from our current vantage point; or we can even discover an object that is not in view, behind us or around a corner. If a Rubik's Cube sits in front of a mirror, for example, we can now see the rear face of the cube.

Shadows too can give access to the presence of things unseen and to the hidden aspects of visible things (chap. 1 and fig. 4.8): we can notice both the presence of a hidden object and some of its properties (if I face a person, I may not be able to judge the profile of her nose, but it is revealed by her shadow on the wall). However, all else being equal, reflections provide a much richer access than shadows, as they preserve colors and internal structure. To put it simply, we would rarely be fooled by a cast shadow to the point of mistaking it for the object that casts it—objects seen in silhouette are the only plausible candidates for the mistake. On the other hand, the informational fullness of a reflection can easily make us believe that we are in front of the real thing. Thus:

Reflections, like shadows, provide access to a scene from a different viewpoint and include, unlike shadows, information about color and texture of objects.

The second advantage of mirrors is the ability to navigate through the world from a different viewpoint. If we use a periscope, we not only see things above the water's surface that we cannot see from our position underwater; we can also move through the environment as if we were at the position suggested by the mirror. We can use the rearview mirror of our car not only for noticing vehicles behind us but also for driving in reverse in a narrow alley. This kind of optical telepresence is being replaced with electronic video versions in many situations, such as backup cameras in cars or remote control video for robots and drones.

Shadows too provide alternate viewpoint access—they provide the viewpoint of the light source, although it is relatively unusual to be able to use that information for action and navigation purposes. An example would be moving around the light source and using shadows to control our movement (fig. 9.6).

Figure 9.6
Taking the light's viewpoint. The shadow actress controls her movement of the light
source by looking at the shadows of the obstacles. Actress: Francesca Bizzarri. Image
credit: RC, Paris, 2002.

9.2.2   Mirrorness and the Visual Logic of Mirrors

9.2.2.1   Mirrorness   Reflections have a rich structure providing scene and
layout information, but how much of it does our visual system use, and
according to what rules? We have seen how the informational structure
of shadows (chaps. 1–2) may or may be not exploited by the visual system
(chaps. 4–5) and how shadow labeling is subject to the presence of mandatory properties (chap. 6). In this section, we discuss mirrorness (what makes
mirrors appear as mirrors, and how painters create the impression of mirrorness), the properties that link a reflection to its source object, and the
logic the visual system uses in dealing with reflections.

To use a visual item as a reflection, we must first label it as a reflection; this entails that a certain surface should be labeled as a mirror. Much
as there is a difference between being a shadow and being perceived as a
shadow, so there is a misalignment between being a physical mirror and
mirrorness, that is, the impression that we see a mirror: not all reflections are
seen as reflections (fig. 9.7), and we can convincingly convey the impression of mirrorness by using nonreflecting surfaces (later we will see the
same dissociation with transparency).

What, then, are the conditions for mirrorness, or mirror character? As
figure 9.7 (left) shows, the image of the mirroring surface should not be
uniform in color or brightness. Moreover, mirrorness is related to "windowness" (fig. 9.7, right) but does not collapse unto it. The content displayed
in the mirror must be (conceptually or visually) relatable to the content

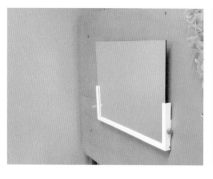

Figure 9.7

Mirrorness. *Left*: To look like a mirror, being a mirror is not enough. *Right*: A change in perspective reveals that the surface is reflecting details of the room to the left (or perhaps it is a window open onto another room?). Image credit: RC, Monza, February 2016.

displayed outside the mirror. For instance, disrupting symmetries can destroy mirror character (fig. 9.8). Recognition of mirrorness should induce discounting of the content of the mirror area in the image. Reflections are not treated as objects (fig. 9.8, left) unless we are unable to attribute mirrorness to the area containing them.

Mirrors need content to appear as mirrors. (But not all content will do.)

Horizontal planes (water bodies) are the predominant, possibly unique case of natural reflecting surfaces. This appears to have been registered in the evolution of vision (e.g., by systematically discounting reflections on still water). In contrast, nonhorizontal mirrors have no evolutionary history and should pose perceptual or interpretive problems. For example, in figure 9.9 we may expect the vertical item—the chimney—to connect to its reflection as a solid, straight line.

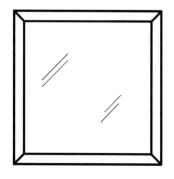

Given their evolutionary novelty, vertical mirrors require extra cues to signal that they are a mirror, showing reflections, a fact noticed by painters. In line drawings, mirrors, like clear windows, are often signaled by oblique lines. More realistically, painters frequently include small highlights at the borders of a represented mirror to indicate the presence of the reflecting surface. Beyond these supporting touches, the critical cue to the reflection is the symmetry between an object and its copy in the mirror space.

Symmetry establishes ownership between an object and its reflection.

Figure 9.8
Violation of symmetry destroys mirrorness. *Right*: In the modified picture, we have the impression of seeing colored, textured patches belonging to the surface. *Left*: In the unmodified photograph, patterns on water are seen as reflections and discounted. Image credit: Cavanagh, Chao, and Wang (2008).

Figure 9.9
Do we (erroneously) expect reflections of vertical objects to themselves be vertical? Image credit: RC, Etna, New Hampshire, October 2013.

However, the symmetry is not mandatory, as a mirror may be reflecting parts of the scene that we cannot see directly. As a consequence, we are at times surprised when we discover that the surface we are looking at is not a mirror but a window. A frame on a window can convey the impression that there is a mirror in its place. We approach the surface from an angle, and it does look like a mirror. But when we move in front of it, we do not see our reflection in it! We do not see ourselves in paintings of mirrors, either, but that does not make them look less reflective, so the cues to mirrorness are quite flexible.

Mirrors provide a three-dimensional space beyond their surface. If the scene around the mirror does not provide for that space, then the surface is more likely a mirror than a window.

People who have mirror agnosia do not understand the workings of mirrors and reach into the mirror to retrieve things they see there. This is a pathological condition associated with right parietal lesions. On occasion, normal vision too is fooled by mirrors. If the mirror surface is not normal to the direction of sight, we do not have the reality check of our reflection and can make an incorrect assessment of the three-dimensional display. The reality check itself is the likely result of development, as it sets in at eighteen to twenty-four months of age: before that time, children are unable to recognize that what they see in the mirror is their own image.

Taking mirrors for windows, and conversely windows for mirrors, is possible because of the high visual expectations we have of mirrors. Reflections are so visually rich that we can be uncertain about windowness versus mirrorness.

Mistaking shadows for real objects only happens when we look at silhouettes cast on a screen, such as a bedsheet drying in the sun, interposed between us and the light source, or a silkscreen in a traditional Edo pavilion (fig. 9.10). The screen abundantly filters light and thus lowers our visual expectations for objects, making it possible for silhouettes to compete with them.

9.2.2.2 Tolerance for errors in content and location: The Venus effect and copycat reflections   Pictorial representations of mirrors bear witness to the fact that we are systematically mistaken in interpreting reflections. On the one hand, we may not accept perfectly correct reflections, typically

Figure 9.10

Miyagawa Chōshun (1682–1752), *Genre Scenes in Edo* (detail), early eighteenth century. Hanging scroll, ink on paper. Idemitsu Museum of Arts, Tokyo.

empty mirrors where we erroneously expect them to host the reflection of an object. On the other hand, we tolerate certain impossible reflections— much as we tolerate certain incorrect shadows (fig. 9.11 and opening figure of this chapter, both digitally manipulated). Impossible reflections of people in mirrors are perfectly acceptable. One particular example, the "Venus effect," summarizes our tolerance for misplaced mirror images. An entire genre of paintings depicts Venus checking herself in the mirror, a setting that we replicate in unretouched photos in figure 9.12. When painters depict people looking at themselves in a mirror, they invariably show the person's face in the mirror. They do not want to show the correct image of an empty mirror, as this would not look like a mirror and thus spoil the whole point of the painting (fig. 9.12, right). But if we see Venus's reflection in the mirror, then geometry tells us that she must be looking at us, not herself. And yet we do not notice anything wrong; it appears that she is looking at herself. More generally, people also have little idea of what to expect in real mirrors.

Reflections do not have to match the reflected scene to be effective.

Although we are relatively tolerant of incorrect locations of reflected objects in vertical mirrors, mirrorness for a horizontal surface does require some semblance of the vertical symmetry that always holds in reality between the source object and its reflection. Destroying the symmetry often destroys the

Figure 9.11
Who ate the croissant and drank the coffee? Reflections do not have to match the reflected scene to be effective. Image credit: PC, RC.

reflective quality of the surface (recall fig. 9.8), and painters are extremely aware of this requirement for reflecting water surfaces. It is difficult to find an example of a deviation from vertical symmetry. The reflection of the moon in the painting by Tosa Mitsuoki (fig. 9.13), for example, is offset from vertical alignment and should not even be visible on the lake when the moon is at such a low angle (the line of sight of the moon reflecting off the water, should be blocked by the intervening hills). The vertical mis-alignment seems wrong in some ill-defined way, but the impossibility of the reflection is not so easily noticed; it requires additional mental geometry.

Figure 9.12

In these unretouched images, the character on the left appears to look at himself in the mirror but in fact does not; instead he is seen by, and can see, the camera. He looks at a point between the mirror and the reflection of the camera to give the impression of the appropriate gaze direction in the reflection. On the right, he appears to stare at a strangely empty mirror, but in fact he now sees himself in the mirror, even though the camera cannot see him. This is known in the literature as the Venus effect. Image credit: PC, RC, Paris, September 2017.

Figure 9.13

Misaligned reflections. Tosa Mitsuoki (1617–1691), *Lake Biwa*. Image credit: Harvard Art Museums/Arthur M. Sackler Museum, Bequest of the Hofer Collection of the Arts of Asia, 1985.352.56.

Figure 9.14
Left: Which mug owns the reflection? *Right*: Isn't something missing in the reflection? Image credit: RC.

Another analogy with shadows is the use of a copycat solution in solving the problem of reflection ownership. In figure 9.14, it takes a little thought, and some knowledge of mirrors' workings, to rule out the first ownership that comes to mind.

Copycat rules, or infringements of those rules, bias the solution of the reflection ownership problem.

9.2.3 Interactions of Shadows and Reflections

9.2.3.1 Shadows on mirrors: Making the difference visible Shadow images are visible on the projection surface, whereas mirror reflections appear to be in a different location, beyond the mirror's surface in the case of plane mirrors (see again fig. 9.2). What happens to shadows that fall on a reflecting surface? In a perfect mirror, all light is reflected at an angle identical to its angle of arrival, so its surface is invisible, and a shadow cast on its surface ends up being visible in the virtual space, and not at all on the mirror's surface. However, not all mirrors are perfect, and scattering from surface dust or muddy water will render shadows visible by diffusing some of the light in all directions. When shadows are visible, they appear to be the same as ordinary shadows cast on nonmirroring surfaces. In figure 9.15, a shadow on muddy water can be compared to a reflection projected by the same object (a tower) that casts the shadow. The imperfect reflecting qualities of the surface may have the further consequence of making the mirror

Figure 9.15
Different geometric behavior of shadows and reflections. The reflection of a point by
a horizontal plane always lies along a vertical line through the point; note, for example, the small chimney and its reflection. Shadows are not so constrained. Image
credit: RC, Chevrainvilliers, September 2009.

image appear to "adhere" to the surface rather than appearing in depth
below the actual tower—here possibly because the tower's features are too
indistinct in the reflection to establish the vertical symmetry.

9.2.3.2 Reflections, shadows, and viewpoint

Figure 9.16 on the left dis-
plays the reflection of a hand in the mirror showing a left–right reversal—
the typical mirror symmetry we expect. We see the same symmetry in the
shadow of a hand held at the same orientation to the wall in the right-hand

Figure 9.16
Mirror and translation symmetries can be seen both in reflections and in shadows. Image credit: PC.

image. In both cases, when the hand is mostly perpendicular to the mirror or shadowed surface, the result is mirror symmetry of the hand's profile. The differences are interesting as well. The mirrored hand is smaller than the actual hand because it seen farther away; in contrast, the shadowed hand is similar in size to the actual hand. In addition, the mirror reveals the hand's internal detail, so we know that we are seeing the palm of both the real and the virtual hands. The shadow of the hand is only the profile of its silhouette, so it has no front or back. We can imagine that since the light sees the back of the hand, the shadow on the wall represents notionally the back.

When the hand is held parallel to the reflecting or shadowed surface, we now see translation symmetries in both cases, again limited to the object profiles. The mirror continues to show the palm of the hand, whereas now it is the back of the hand that we see in direct view, so the translation is of the profile only, not the internal detail. The shadow shows a translation of the profile of the hand, but without any internal detail, so like the mirror, it does not show translation symmetry for internal detail.

This means that an object needs to be relatively flat and reasonably parallel to the mirror or shadowed surface to make copycat shadows or copycat reflections (obtained by translational symmetry; see also chap. 4). Only when the object is more perpendicular to the mirror or shadowed surface do we see mirror symmetry (of the profile).

In translational symmetry, the shape of the object is copied onto its image by translating (i.e., shifting) each point of the object by the same given distance along the same direction. For instance, the cursor of a word processor translates on the screen when you press the space bar or the arrow keys. In a mirror symmetry, the shape of the object is copied onto its image by copying each point at the same distance from a given symmetry plane (the mirror surface). So mirrors always have bilateral symmetry, and only in the special cases we have described do we see translational symmetry for the profiles (while ignoring inner details).

We now introduce two pictorial examples of viewpoint impossibilities related to mirror reflections and shadow projection. The error in viewpoint is obvious in Magritte's *La reproduction interdite* (and other artists' variations on it; fig. 9.17). We should see the face of the person in the depicted mirror (assuming, and the reflection of the book supports the assumption that

Figure 9.17
This is not a mirror. Elia Grop, *La reproduction interdite de la reproduction interdite, d'après R. Magritte*, 2017. Liceo Artistico Nanni Valentini di Monza, Laboratorio audiovisivo multimediale. Image credit: Elia Grop.

Figure 9.18
This is not a real shadow. As we are on the same side of the statue as the light source, the shadow of the statue should not be mirror inverted in the image. Redrawn from Giorgio De Chirico (1888–1978), *Piazza (Souvenir d'Italie)*, 1925. Private collection. Image credit: RC.

the character is facing a mirror), and not a copycat replica of what we see of him from our viewpoint. The mirror has retained translational profile symmetry but has shown the internal detail from the back rather than the front as it should.

Giorgio De Chirico's *Piazza* (fig. 9.18) offers a case of mirror reflection of a shadow that violates the expected translational symmetry. Because the observer is on the same side of the statue as the light source, the shadow must be a translation of the statue's image, whereas the picture presents a reflection instead.

9.2.3.3 Shadows or reflections? The example in figure 9.19—a digitally doctored image—shows yet another instance of impaired understanding of reflections and shadows. Are these shadows or reflections? They are neither. Shadows cannot be cast in this way—with such a high contrast—on a

Figure 9.19
These cannot be shadows and cannot be reflections either. Image credit: Advertisement for Fortis Bank, 2009.

perfectly reflecting surface (cf. fig. 9.22). And reflections of a vertical object (here, a human body) on a horizontal plane cannot be misaligned with the direction of the reflected body: the figures in water cannot be reflections of the two hikers, as reflections on a horizontal plane would have to align vertically with the objects being reflected. Moreover, if the figures are interpreted as shadows, we should ask where the reflections of the two walkers have gone. What happened here? Various hints show that the illustrator fabricated the image: the silhouettes of the two walkers are identical, and the lengths of the "shadows" are impossibly different; it appears as if the manipulator wanted to be confined within the picture's frame. But, as in many other cases, it takes a while to discover these impossibilities, which without our scrutiny may have gone forever unnoticed. The visual system does not appear to compute directly the impossibility of the illustration; some step-by-step reasoning appears to be required to ascertain that the illustration is indeed wrong.

9.2.3.4 Mistaking reflections for shadows

If some correct reflections are visually rejected—for instance, if an expected alignment is not respected (fig. 9.9, replicated in the margin here)—they can be relabeled, in a second-best choice, as shadows, in particular if their internal content is diminished and they acquire a silhouette-like aspect. The pitched metallic roof has an odd orientation for a reflecting surface (standard natural reflecting surfaces are horizontal). The visual reasoning seems to go as follows: "This cannot be a reflection because it isn't vertically aligned with the chimney; hence it must be a shadow."

Why do some reflections look like shadows? When a reflecting surface also adds some diffusely reflected light, the contrast in the reflected object is less than the contrast in the object viewed directly. In the image, the area of the reflection loses texture and structure, the same loss that is one of the key aspects of shadow character (shadows of opaque objects have no internal structure.)

Degraded reflections may lose internal structure and appear to be shadows.

9.2.3.5 Mistaking shadows for reflections

All of this said, cast shadows and mirror reflections are informationally close enough to explain why artists sometimes use the geometry of reflections to depict shadows, with occasional success. Japanese woodcut art, for example, did not capitalize on post-Renaissance perspective and shadow depiction, except in some sporadic cases. Japanese artists discovered representational shortcuts for shadow representation (possibly because they did not have the canonical set of pictorial shortcuts at their disposal).

In Hiroshige's *Night View* (fig. 9.20), the "shadows"—this is particularly evident in the rightmost characters—are not properly aligned with the light source casting them (the moon) and are instead aligned vertically with the objects that cast them, thereby suggesting reflections. However, the dark zones are not appropriate for reflections, either, because they are too short.

It is the "times two" regularity that specifies that the "reflections" are too short in figure 9.20. If properly depicted as reflections, they would be the same height as the people. Indeed, in some conditions, reflections may be misattributed, and the distance function then places the original object in an erroneous location. Figure 9.21 shows an example of this "reflection capture" where the island seems to float above the water on poles.

9.2.3.6 Casting shadows through the looking glass What happens, then, when shadows and reflections are combined? And how much can we understand of the resulting situation? In *The New Ambidextrous Universe*, Martin Gardner wrote about the surprising effect created by projecting a light beam in a mirror:

If you want to puzzle and fascinate a small child, stand him in front of a large wall mirror at night, in a dark room, and hand him a flashlight. When he shines the flashlight into the mirror the beam goes straight into the room behind the glass and illuminates any object toward which he aims it.

The mirror reflects the beam of light back into our room, but we see in the mirror the result of this reflection, that is, a beam of light that searches inside the space of the mirror. The effect is surprising because one has the impression of the light beam physically entering the space in the mirror, which is otherwise inaccessible. This invites the idea of generating a similar effect through the projection of a cast shadow into the mirror space by illuminating an object situated on the observer's side of the mirror (fig. 9.22).

There is, of course, nothing mysterious here. The mirror is reflecting light onto the wall behind the viewer. The hand blocks the light, and a shadow forms on the wall behind the viewer. The shadow is visible in the mirror because it is on a wall that is visible in the mirror. But the effect of

Figure 9.20
Utagawa Hiroshige (Ando) (1797–1858), *Saruwaka-machi Yoru no Kei* (*Night view of Saruwaka-machi*), no. 90, 1856. From the series *One Hundred Famous Views of Edo*. Image credit: Toyama and Naitu.

Figure 9.21
Reflection capture. The island appears to be propped up on poles above the lake. What looks like a reflection of the island is actually a reflection of an arbor of wisteria and its supporting poles. Image credit: Seiko Hoshi, Tokyo, Hama rikyu Gardens, 2016.

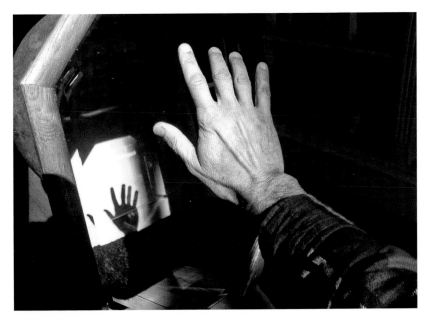

Figure 9.22
A shadow that appears to be cast into the space of a mirror. Image credit: RC, Etna, New Hampshire, March 2014.

Figure 9.23
The camera flash is very close to the camera lens, so it casts a shadow of the teapot into the mirrored space, where it is hidden from our view by the teapot itself. Nevertheless, light from the reflected light source is blocked by the virtual teapot, and it casts a shadow on the wall on our side of the mirror. The shadow that we see is equivalent to the shadow cast by the virtual teapot, back into the real scene! This visible shadow cannot be seen as a reflection in the mirror because it is the same cast shadow that is blocked by the teapot. Image credit: RC, January 2015.

projecting something *into* the space of the mirror is compelling, especially if we wave our hand.

In some cases, shadows appear to be cast from the real world into the virtual world of reflections.

Consider figure 9.23, where the light source and the point of view coincide, so that shadows produced directly by the light source are not visible. Nevertheless, on "our" side of the mirror, we see a teapot shadow that is equivalent to the shadow of the virtual teapot cast by the virtual light source. Intriguingly, this shadow is not visible in the mirror—because it is masked by the teapot itself.

In some cases, shadows appear to be cast back into the real world from the virtual world of reflections.

9.2.3.7 Reflections as objects: Chimeras If shadows can build chimeras (chap. 6), so can reflections. The simple photo trick used in figure 9.24 tells an interesting story: we compute the object despite knowing that half of it is just a reflection and although the resulting chimera is too thin, has divergent eyes, and so on. The video version of this image, with the person "flying" for a significant amount of time, further tells us that we do not care about the biology (people cannot levitate).

9.2.4 Common Computational Resources for Reflections and Shadows

Shadows and reflections share a few computations. In the instances where they both involve a source object and its shadow or its reflection, both give rise to determinations of ownership and relative position. For example, it is true in both cases that if an object does not touch its shadow or its reflection in an image, then the object does not touch the surface on which the shadow falls or the surface making the reflection. We can identify many other differences, but it would be plausible to say the following:

Shadow computations may share computational resources with reflection computations.

Further Research Question: Reflections

1. It would be fascinating to dig deeper into the computational relationships between reflections and shadows. For example, recall the patients with mirror agnosia described in section 9.2.2.1. They take the mirror space as containing real objects rather than reflections, and they reach for them. Would they also have difficulty with shadows, mistaking them for objects? Developmental studies could provide another source of comparison. Does the use of shadows for depth perception emerge at the same or a different time from that for the perception of reflections? We do have many studies on the emergence of shadows as a cue to depth and spatial layout in infants (around six months). The studies of self-awareness in mirrors show that this emerges later (around 2 years). Any correlation between the age of emergence of these two abilities would suggest a relation, but not evidence of causality. New studies of infants' use of mirrors to locate objects and the emergence of the understanding that reflections are not objects would give us more information, but these studies are not available yet.

Figure 9.24
Flying chimerical character. Image credit: RC, Paris, September 2016.

9.3   Transparency

9.3.1   The Ecology of Transparency

If shadow perception has various aspects in common with the perception of reflections, it has a lot in common with the perception of transparency as well. In particular, both share the critical role of X-junctions and contrast relations at the border of the shadow or transparent material (fig. 9.25). Researchers noted this similarity a long time ago, but a clear assessment of the similarities and the differences has been possible only after the rise of computational accounts of transparency. Fabio Metelli observed that physical transparency and phenomenal transparency are decoupled: perfectly transparent panes of glass do not appear transparent—they do not appear to have any presence at all until we bump into them. Conversely, the impression of transparency can be conveyed by opaque surfaces with appropriate luminance relations. Following the path used for shadows and reflections, we can distinguish three aspects of transparency: information in physically transparent structures; the uses of transparency in further visual computations (mission); and the properties required for the appearance of transparency (what we may call transparency labeling and character or "transparenciness," which Metelli called "phenomenal transparency").

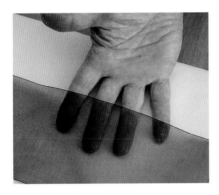 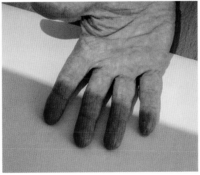

Figure 9.25

Left: The border of the transparent material makes X-junctions with the contours of the hand. *Right*: Similarly, the shadow makes X-junctions with the contours of the hand. Image credit: PC.

Figure 9.26
A faulty glass is seen as a transparent medium in front of a background. Image credit: RC, Trieste, April 2016.

Near-perfect transparency, like that of the atmosphere on a clear day, is useful but not very interesting visually. Solid objects with nearly perfect transparency were manufactured only in recent times, when technology made possible the production of faultless glass (fig. 9.26). In our polished-window-rich world, we have to use stickers on window panes to prevent birds from bumping into them, suggesting that the avian brain is as blind to perfectly transparent objects as we are (fig. 9.27). Of course, a clear atmosphere or body of clear, still water is transparent as well, but we will exclude these examples from our discussion for the moment.

9.3.1.1 Information and mission   A transparent material is any material that lets some light through—in other words, it is not completely opaque. This covers an enormous range of materials, including some that may distort, diffuse, color, and attenuate the light as it passes through, and others that are optically clear, like a perfect glass. Here we will focus on the classic

Figure 9.27
Stickers are needed to signal the presence of a well-cleaned glass. Image credit: RC, Paris, September 2017.

case of a transparent film lying over a background surface (fig. 9.28). The light may be reflected diffusely from the front surface of the material, adding a constant amount of light everywhere. The remaining light then passes twice through the film, once directly through (a translucent material would add some optical scattering here) and once again after diffuse reflection from the background. So if the transparent material lets through 50 percent of the light that strikes it, only 25 percent of the incident light returns from the double voyage through if the background surface is a pure white, and less if the background is darker (fig. 9.28). The information revealing these

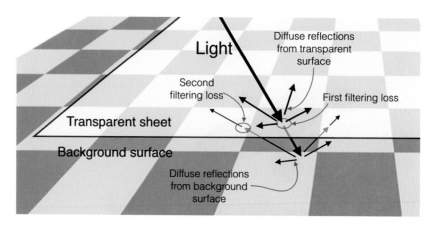

Figure 9.28

Light passes twice through a transparent film lying over a background (depicted here with a slight separation for illustration purposes). Any diffuse surface reflection will add a constant amount of light everywhere. The light diffusely reflected from the background surface makes it back through the transparent layer and to the viewer. Image credit: PC.

effects of the transparent surface comes from the comparison of luminance and contrast changes across the border of the material (fig. 9.29).

Transparent materials may both filter light and reflect light.

The mission in seeing transparency is to factor out these effects of the overlying material and recover the material properties of the underlying surface—how it would be seen in the absence of the transparent overlay. This is similar to the discounting of the illuminant process for shadows, but now it has two dimensions to correct: luminance, like a shadow, but also contrast.

Transparency does not have the dual relation seen for both shadows and reflections, where an object and its shadow or an object and its reflection could offer information about the spatial layout of the scene. Nevertheless, transparency does come with a sense of depth ordering—that the transparent material lies in front of the surface behind it. But unlike shadows and reflections, transparency does not offer any further information about the spatial layout of the scene. Also, unlike shadows and reflections, the transparent material is not demoted; it is an object in the scene. As yet, no studies have compared the speed of finding an oddly oriented transparent

Figure 9.29
The transparent sheets reduce the contrast of the book cover below because of the light being reflected diffusely from its surface. The cues to the transparency are the X-junctions where the edges of the filters cross each other and the letters and the levels of light, gray to black, in the four regions around the central X-junction. Image credit: PC.

material, but a sample search array suggests that an oddly tilted transparent bar should be found as quickly as an oddly tilted opaque bar (fig. 9.30).

9.3.2 Logic of Transparency

9.3.2.1 Mandatory properties: Metelli's rules The perception of transparency has been a topic of scientific research for almost 150 years. The most significant contribution to the study of transparency has been a set of luminance relations described by Fabio Metelli. These are now called Metelli's Laws, referring to the luminance relations between parts of a surface viewed directly and those parts of the surface that are seen through a transparent layer. In addition to transmitting light, a transparent material may also reflect some light (a material that both reflects and transmits light both direcly and diffusely is sometimes considered translucent; fig. 9.35). This represents an extremely small subset of the natural occurrences of

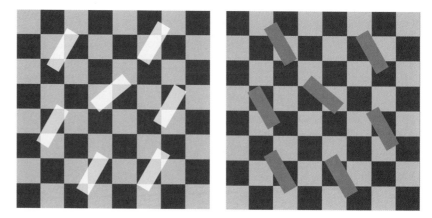

Figure 9.30
On the left, all items are transparent bars. The search for the odd orientation seems not much more difficult than on the right, where the bars are opaque. Image credit: PC.

Figure 9.31
Transparency can be seen in the left and center examples, but not on the right. The left-hand example is bistable; either the top or bottom square can be seen as transparent, but not both at the same time. In the middle example, only the bottom square can be seen as transparent. The luminance relations that govern these patterns are described in the next figure. Image credit: PC.

transparent materials, some of which we mention at the end of this section. It does, however, present an easily analyzed and fairly well-understood case study for transparency.

These Metelli relations are depicted in figure 9.31. The regions where the surfaces cross create X-junctions, and the luminance relations between the four parts making up the X-junction are closely linked to the perception of transparency and the corresponding depth order. On the left, transparency is perceived: the transparent layer has reduced the amount of light returning from the region where it overlaps the opaque gray square on the background. This darkening is symmetrical, and either square can be seen as transparent, although not both at the same time. The percept is bistable, and when the square that appears transparent switches, so does the depth order.

In the middle example, the lower square has two effects: it reduces the amount of light returning from the background, but it also adds some light that is diffusely reflected from its surface, as a sheer curtain would. So the lower square is darker than the white background but its surface reflection lightens the overlapped region. Now there is no ambiguity in regard to the depth order of the two squares: the lighter lower square appears in front of the darker square. On the right, no transparency is perceived, and we see the three regions as a flat mosaic. (Note that, in all three examples, talented observers may also perceive one of the surfaces covering the entire image with a hole through which the other is partially seen.)

These three relations are quite tolerant of luminance variations within the bounds that define each, but they also have well-defined criteria (fig. 9.32). The case on the left has the same direction of contrast change along all borders. Along the horizontal border of the top left X-junction, the luminance changes from darker to lighter across the border and does so on both sides of the vertical boundary. The same is true for the vertical border: crossing it from right to left, the luminance changes from darker to lighter on both sides of the horizontal boundary. This is not true in the middle figure. Here only the vertical border of the left X-junction has this property. Along the horizontal border, the polarity reverses from dark to light on the left to lighter to darker on the right. This reversal is due to the

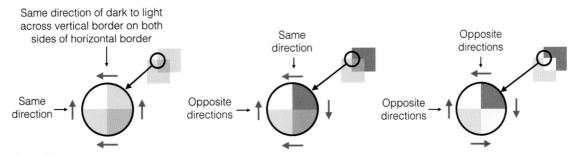

Figure 9.32
Contrast polarity order at the X-junctions defines the three interpretations seen in fig. 9.31. The left-hand X-junction of each pattern is isolated, and the order of light-to-dark changes across the border is noted. There are no direction reversals on the left, one in the middle, and two on the right. Image credit: PC.

extra luminance added by the surface reflection and unambiguously specifies which square is the transparent one. On the right, the polarity reverses along both borders, and there is no possible transparent interpretation for this combination.

The change from light to dark of a background material must have the same direction when seen through the transparent layer.

The broad tolerance for quantitative variations within the three zones of interpretation arises because the transparent layer has two degrees of freedom: the proportion of light transmitted, and the proportion of light diffusely reflected from its front surface. As the luminances seen through the transparent layer vary, the perception of the transparent material changes to give an appropriate balance of transmitted and reflected light that matches the luminance values.

When the transparent material has a significant diffuse surface reflection, it adds an equal amount of light to all the regions of the underlying surface. This reduces the contrast seen between adjacent parts of the rear surface. We see this effect in figure 9.33 on the left, where the luminance difference between the black and white in the background is much larger than the luminance difference between the same two background regions seen though the central gray disc. The importance of the X-junctions for establishing transparency can be seen on the right in figure 9.33, where

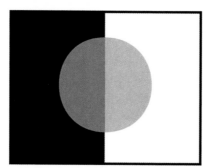 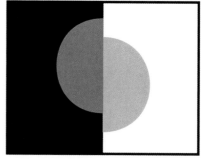

Figure 9.33
Left: The transparent material here produces a lower contrast for a background edge where it is seen through the material. This indicates that the material is also reflecting light diffusely from its surface, adding an equal amount of light to both sides of the underlying border, reducing its contrast. *Right*: A demonstration that the X-junction in the left image is critical to the impression of transparency. Image credit: PC.

turning the X-junctions into T-junctions by shifting the two halves of the circular layer eliminates the transparency percept. (However, explicit X-junctions are not always necessary for the perception of transparency.)

9.3.2.2 Beyond Metelli's rules The classic literature on transparency focuses on the role of luminance patterns of flat surfaces and their X-junctions, but our everyday experience is more frequently with glass or plastic objects, puddled water, thin sheer curtains or clothes, or translucent barriers (figs. 9.34, 9.35). These cases offer many more cues to the presence of a transparent material than those dealt with by the Metelli constraints. Highlights and shadows and optical effects establish the properties of the material, as well as the similarity of the background seen through the material compared to that around it. When an object is entirely behind a partially transparent material (fig. 9.35), there are no X-junctions or comparisons of the object properties and textures directly viewed versus viewed through the material. Here, if the visible pattern is recognizable as a familiar object or structure, the deviation between the pattern and its canonical form can be attributed to degradation from a less than perfectly transparent material. If the degradation is extreme or the object

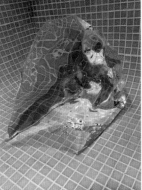

Figure 9.34
Everyday transparent materials like a glass (*left*) and a block of ice (*right*) offer many cues to their transparency beyond X-junctions and Metelli luminance properties: highlight, shadowing, and optical properties, and similarity of background texture seen through the material compared to being viewed directly in the surround. Image credits: Left: PC. Right: RC.

Figure 9.35
When objects are viewed through a translucent glass, the deviations from the object's known form are attributed to imperfections of the glass medium. Image credit: PC and RC.

unrecognizable, we may instead see the pattern as part of the material's surface texture.

Recent studies in the field of material perception have examined the role of other properties of transparent materials, such as lensing effects due to refractive index changes. Researchers have also examined the color constraints for transparency, and findings suggest that we are not very good at discounting the color of the overlying filter (fig. 9.36). This is in contrast to the luminance effects of overlying filters or of shadows, where the recovery of surface reflectance (lightness) is quite good.

Highlights, hatching, superimposed textures (referred to as "laciness" by Watanabe and Cavanagh), and object knowledge have provided effective means for representing transparency outside the realm of the Metelli rules. We know this because artists have been trying to convey transparency for centuries (fig. 9.37).

There are many ways to convey transparency. Some are simply not available for shadowness.

Figure 9.36

When a red transparent sheet is placed over a background pattern, it filters out colors other than red and also adds some diffuse red scattered light from its surface. This makes the white parts of the underlying surface indistinguishable from the red parts; there is no discounting of the filter's color properties. This differs from the discounting (called lightness constancy) seen for the luminance changes in shadows and those for neutral transparent materials as well. Image credit: PC.

Figure 9.37

Left: The similarity of objects seen above and through the bowl indicates the bowl has some transparency. *Right*: The transparent water is also conveyed by the similarity of the darkness seen in and around the glass. Image credits: Left: Fresco in the Villa di Poppea in Oplontis (ca. 70), photograph Lise Hannestad, 1979. Right: Jean Siméon Chardin, *Water Glass and Jug*, 1760 (detail), Web Gallery of Art.

9.3.2.3 Properties of transparency that are ignored by the visual system

It is not clear that vision ignores any of the basic Metelli rules for transparency. Transparency was an early addition to realism in art, dating to Egyptian times (fig. 9.38, left); and from that time on, painters got the local properties of transparency mostly correct—that is, the adjacent luminance values trigger the perception of transparency. Across several thousand years of artistic experiments with the depiction of transparency, we do not find examples that violate the luminance properties of transparency but are nevertheless seen as transparent. In this respect, this constraint is similar to the basic darkness property for shadows and the vertical symmetry property for reflections on still water surfaces: artists break these at their peril. The depiction no longer captures the intended shadow or reflection or transparency character.

Failures of transparency are occasionally seen in art (fig. 9.38, middle), but even during periods when the depiction of cast shadows degenerated and then disappeared, transparency remained relatively well depicted (fig. 9.38, right). The point here is that bad transparency is failed transparency,

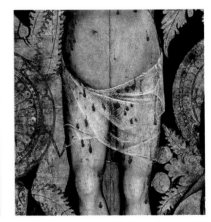

Figure 9.38

Left: Egyptian artists captured transparent materials extremely convincingly using reduced contrast and X-junctions. *Middle*: Some Roman attempts to depict transparency were not so successful. The water here has a uniform effect on contrast, as if it were a veil, rather than a graded effect. *Right*: Even during the period when Western artists did not depict cast shadows (400–1425 CE), transparency continued to be rendered realistically. Image credits: Left: Tomb painting of Nefertari (Nefertari Merytmut or Mut-Nefertari) (ca. 1290–1255 BC), wife of Ramses II (Ramses the Great), Valley of the Queens, near Luxor, Egypt. Photograph, Brian Brake/Science Photo Library. Middle: Villa Romana del Casale, Piazza Armerina, mosaic of capture of rhino, early fourth century. Ivan Vdovin, Alamy Stock photo. Right: Juan Sanchez, *La Crucifixión* (detail), ca. 1310. Museo de Prado, Madrid.

but an absence of cast shadows is not particularly harmful; after all, there are many cloudy days when we cast no shadows.

However, painters have often violated the refraction index or color filtering properties of transparent materials without any loss of transparency—these properties do not seem to be verified by the visual system. In fact, colored filters act by subtracting color, and the visual system may not be very good at coping with this. This is understandable because the subtractive relations between two colors cannot be predicted from their perceived colors; these relations depend on the colors' respective wavelength composition—two reds that look identical to a human observer may have quite different spectral compositions and so act differently in a subtractive case (e.g., mixing pigments or superposing colored filters). Indeed, one study showed that when colored words were superimposed, the recognition of the words was best when the overlapped regions had additive color relations, as would overlapping colored lights. So where blue and yellow overlapped, the best recognition occurred when the overlapped regions were white (the additive mixture), not green (one possible subtractive mixture). Similarly, the refracting properties of glass or water are not easy to compute, and the visual system may simply allow a range of distortions or none at all as the possible lensing effects of a transparent material.

Violations of the luminance properties of transparency generally destroy transparency, but color and optical properties can be violated.

9.3.3   Similarities and Differences with Shadows

9.3.3.1   Commonalities
For the study of the similarities between shadow and transparency perception, the Metelli rules are the crucial ones; they appear to be a superset of the rules for shadow labeling and character. X-junctions are critical in both cases, as is the pattern of luminance across boundaries. Recall that a key condition for the acceptance of shadows is the preservation of luminance ratios across the shadow boundaries (fig 6.16). The ratio had to be less than one (shadows are darker) and similar at all points along the border.

This particular pattern of luminance change across the border also holds true for some transparent materials (those that have pure attenuation, no surface reflectance). We see this effect on the left in the shadow example in figure 9.39, as well as on the left in the transparency example shown

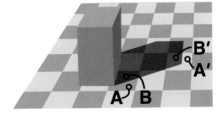

Ratio of contrast preserved, B':A' = B:A, for shadows and for transparency.

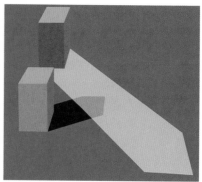

Figure 9.39

Left: Two blocks lit separately from different directions produce shadows that superimpose with X-junctions and luminance relations appropriate for either shadows or transparency. *Right*: A shadow superimposed on a transparent material that seems to cover the shadow. The luminance pattern may be appropriate for transparency, but not for shadow. Image credit: PC.

earlier in figure 9.31. Shadows can only subtract light, whereas a transparent material can both subtract light coming from the background and add light because of its own surface reflectance. So, in figure 9.39, the example on the right supports transparency for the lighter region, but that region cannot be seen as a shadow. In this case, light is added from the transparent surface, a situation that has no equivalent for shadows.

9.3.3.2 Resilience of transparency to contour lines

On top of the perceived and computational commonalities, there are other important differences; these may depend on the fact that although they may use the same basic computational resources, different sets of rules play a hand. For example, we know (from chap. 4) that outlining a shadow region destroys its shadow character. In contrast:

Contour lines do not automatically destroy transparency.

Contour lines cannot be interpreted as boundaries of illumination, whereas transparent sheets are full-blown objects and can have their own bounding contour (fig. 9.40).

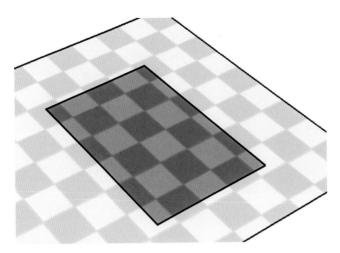

Figure 9.40
Outlines do not destroy transparency, although they would destroy shadowness in the same conditions. Both cases preserve X-junctions. Image credit: RC.

9.3.3.3 Shadows are not transparent

If transparency conditions hold, we do not necessarily see transparency.

As a case in point, we do not see shadows as transparent. In particular, perceptual ordering of surfaces in depth characterizes transparency (the transparent surface is seen to lie in front of the background) but does not characterize shadows. A strong claim to the contrary was made by Arnheim, according to which shadows are like superposed transparent film: "The superposition observed on the surface of illuminated things is … a transparency effect." But shadows are two-dimensional illumination phenomena, involving a surface and a distribution of light, whereas transparency is tied to objects, involving two superposed surfaces (fig. 9.41).

9.3.3.4 Blurred edges There is more to shadowness than just the satisfaction of a subset of Metelli rules at X-junctions. In particular, among the signature differences, blur of the shadow border is a strong mark (fig. 9.42) for shadowness (or, in general, for illumination differences) but is rare for transparent materials.

Figure 9.41
Shadows do not appear as transparent even when there are two of them. Neither of the two shadows appears to be in front of or behind the other. Image credit: RC, Strasbourg, April 2016.

Figure 9.42
Shadows frequently have blurred edges, whereas transparent materials rarely do. Image credit: PC.

Figure 9.43
The black and white bounding contour diminishes shadow character and enhances the interpretation of a superposed transparent film. Image credit: RC, Turin, March 2016.

As a consequence (fig. 9.43):

If shadow character is lost owing to an inappropriate outline, the shaded area can be seen as a transparent overlaid surface.

9.3.4 Common Computational Resources for Transparency and Shadows

Stoner (1999) suggested that transparency perception relies on the processes of shadow perception:

Given the relative rarity of transparent surfaces in natural scenes, it might seem puzzling that the visual system would devote neural machinery to its detection. An alternative possibility is that perceptual transparency depends on mechanisms that typically process the more common visual phenomena of shadows and opaque occlusion. Shadows and opaque occlusion are ubiquitous in natural scenes and, moreover, can be thought of as defining limiting cases of perceptual transparency.

We agree that, given the overlap in the critical properties for shadows and transparency, it is highly plausible that they are based on common computations.

Shadow computations share computational resources with transparency computations.

It is difficult at this stage of research to claim much more than this, and it would be important to be able to dig deeper into the computational relationships between transparency and shadowness.

9.4 Conclusions

Mirror reflections, like shadows, are projections from an object into a space different from the space of the object. Studies have documented mirror-specific visual computations, in particular tolerance and even preference for incorrect positioning of objects in represented mirrors, as well as intolerance for correct reflections. This invited studying reflections and shadows comparatively. We explored the ecological and informational analogies and differences of shadows and mirrors in a variety of situations in which reflections could assist image and viewpoint retrieval in a manner close to that made possible by shadows. We underscored the extent to which the character of mirrorness echoes the character of shadowness. Both reflections and shadows can be and are used but often must be discounted.

The visual analogy of shadows and some reflections produces some computational failures. In the case of position retrieval, applying to reflections the "times one" rule that holds for shadows creates mistaken assessments of an object's distance from surfaces. In the case of retrieving the viewpoint of a light source, applying to shadows the same rules for reflections deceives us as to which side of the shadowed object we are looking at. The complexity of the interaction of shadows and reflections in cases in which shadows are cast into the space of the mirror may be computationally intractable, so that we are surprised at perceptual outcomes such as missing shadows or shadows that appear to bridge the gap between our space and the space behind the mirror.

If an object does not touch its reflection, it does not touch the reflecting surface, either. This is true of shadows as well.

An object's reflection is always smaller than its shadow. If the shadow is located approximately in the position in which a reflection could be located (i.e., between the observer and the object), then the visual angle of the cross section of the shadow will necessarily be larger (assuming a light source of small extent, like the sun) than that of the corresponding reflection.

A reflecting surface is halfway between the source object and its reflection. This is the times two rule. On the other hand, the distance from an object to its shadow depends on the direction of the light and is only equal to the distance to the surface (times one rule) when the light is normal to the surface.

Reflections, like shadows, provide access to a scene from a different viewpoint and include, unlike shadows, information about the color and texture of objects. For reflections, this viewpoint is that of an eye placed behind the mirror; for shadows, it is that of an eye placed at the light source that produces the shadow.

Mirrors need content to appear as mirrors. A featureless or uniformly textured reflection that fills the mirror will make the mirror appear to be a painted surface with no reflection.

Symmetry establishes ownership between an object and its reflection.

Mirrors provide a three-dimensional space beyond their surface. If the scene around the mirror does not provide for that space, then the surface is more likely a mirror than a window.

Reflections do not have to match the reflected scene to be effective.

Copycat rules, or infringements of those rules, bias the solution of the reflection ownership problem.

Degraded reflections may lose internal structure and appear to be shadows.

In some cases, shadows appear to be cast from the real world into the virtual world of reflections.

In some cases, shadows appear to be cast back into the real world from the virtual world of reflections.

In the case of transparency, the similarity to shadows lies in the importance of X-junctions and pattern of luminance at the X-junction. This leads to an assumption of some common computational resources underlying both. We can nevertheless point to several differences. Shadows typically have a paired relation between the object and its shadow. Transparency is simply a material property revealed by the effects it has on the scene that is visible through it.

Transparent materials may both filter light and reflect light.

The change from light to dark of a background material must have the same direction when seen through the transparent layer.

There are many ways to convey transparency. Some are simply not available for shadowness.

Violations of the luminance properties of transparency generally destroy transparency, but color and optical properties can be violated.

Contour lines do not automatically destroy transparency.

If transparency conditions hold, we do not necessarily see transparency.

If shadow character is lost owing to an inappropriate outline, the shaded area can be seen as a transparent overlaid surface.

Shadow computations share computational resources with transparency computations.

As we mentioned, visual artists throughout history have gone a long way in the (implicit) study of the rules for representing shadows, transparency, reflections, and illumination. The time has come to give a proper place to their discoveries, and this is our topic in the next chapter.

Further Research Questions: Transparency

1. As with the relation between reflections and shadows, evidence from patients with brain damage would be extremely revealing. We are not aware of any patient cases that report a loss of transparency perception or a loss of shadow perception, so for now we cannot address the question of whether these faculties can be lost independently.

2. Developmental studies would also be relevant. Does the perception of shadows emerge at the same or a different time as the perception of transparency? For the moment, we have many studies on the emergence of shadows as a cue to depth and spatial layout and as nonobjects. We do not have the equivalent studies for transparency.

A master of shadow observation—and some complex shadows he depicted. Robert Campin, *Santa Barbara*, 1438. Image credit: Museo del Prado.

10   The Art of the Shadow

10.1   The History of Shadow in Art

The cognitive study of art has blossomed in the wake of the publication of Ernst Gombrich's *Art and Illusion* (1960). The research has been mostly a one-sided affair, focused on the attempt to dovetail the cognitive underpinnings of art production and appreciation. Art is such a complex and intriguing phenomenon, so specifically human, that it comes as no surprise that the cognitive science community has devoted so much effort to understanding it. However, another, less-explored research path also exists, one closer to Gombrich's original idea: mining the artistic corpus construed as a repository of solutions to certain visual problems. Here the study of art meets the study of perception as a window on the functioning of the visual brain.

For a discussion of some methodological problems involved in analyzing an artistic corpus in the framework of research on the visual system, see chapter 11.

Scholars of perception enlist shadows among the pictorial cues used for rendering the third dimension, along with other cues such as occlusion and linear perspective. These cues are called *pictorial* because they can be used in a painting to convey the sense of space and dimension, as opposed to non-pictorial cues such as accommodation of the lenses in the eye and binocular disparity. It is remarkable that the visual brain uses a dozen or more different strategies, kept in delicate balance and mutual interaction, to retrieve distance in the visual scene; most clearly distance is a core component of the interpretation of the scene (as we have seen in chap. 4).

We know that shadows can reveal the existence, shape, and other features of objects, surfaces, and light sources, but did painters take notice, and how?

The short history of the shadow in art can be made very short indeed. First, cast shadows are nowhere to be found in non-Western art and in

Western art before the Greek and Roman period (though shading is present very early on, e.g., in Lascaux cave graffiti). Second, the Romans, especially in the Hellenistic period, experiment wildly with shadows in paintings and in mosaics, at times generating highly conventionalized images with a rich epidemiology. Third, cast shadows disappear again during the Dark Ages (while shading continues to live a rich life of its own). Fourth, starting in the early Renaissance, painters realize the strong added value of shadows for realism and begin a second wave of pictorial experiments (some of which we detail in this chapter). Fifth, the study of geometry canonizes algorithms for calculating shadows that become standard lore in beaux arts curricula and in architectural rendering. Sixth, shadows begin propagating outside Western art, in particular (albeit timidly) in Japanese art. What, of this history, is relevant to visual science?

A caveat on the limits of our endeavor is in order. Much can be said—and has been said—about the fascinating symbolic and narrative uses of shadows. Shadows with a narrative role and a personality are a popular theme in literature and the visual arts. Our work here stops short of describing these fascinating lives of shadows, as we try to stay within the limits of the visual brain and avoid the complex relations with the emotional and the narrative brain. We touched on the interface between the visual system and cognition in chapter 7. We saw that cognition can act as a judge of the shadow percepts created by the visual system, occasionally disagreeing with and sometimes influencing them. The study of the interface between vision and cognition is useful for scholars who investigate the complex symbols and narratives that turn around shadows. Shadows can do some of the adventurous, sometimes malignant, poetic things that have been attributed to them in part because they have the visual properties they have: they move, rebel, hide, refuse to be identified, vanish—all these visual aspects provide fertile ground for complex metaphors and narrations. Shadows are so visually telling that it takes little to move into emotionally tinged narratives. But it is the visual aspects that we primarily deal with here. We start with some basic functions.

We first look at some of the *positioning* functions offered by shadows that are used by painters. We then consider several types of misrepresentations of shadows—shadows doing impossible things—that nevertheless reap a payoff for scene layout and do not look particularly shocking. We finish with an appreciation of some early Renaissance artists who were keenly observant in their depictions of shadows.

10.2 Pictorial Functions of Shadows

10.2.1 Positioning Shadows and Anchoring Shadows

By far the most significant contribution of cast shadows in paintings is a positioning function. We will use the term "positioning shadows" for shadows that specify the relative situation of an object and the surface on which the shadow falls. Among positioning shadows, the "anchoring" version is the most widespread type (fig. 10.1). In the paintings we studied, most shadows originate at the characters' feet, to signal that the depicted characters are located at some specific spot, anchored on the ground (and are not flying above it, as in fig. 10.2).

The anchoring function does not require sophisticated shadow representations; we know that a few marks on the ground, to suggest a darkening of the area at the contact point, are in general enough. This visual appearance corresponds to the short, larval shadows seen at the base of objects in diffuse light. However, shadow representations are not just adaptations of

Figure 10.1
Anchoring shadows firmly locate people and objects on the ground. Sandro Botticelli (1444/5–1510), *The Last Communion of Saint Jerome* (detail), 1490s. New York, Metropolitan Museum of Art, accession number 14.40.642.

Figure 10.2
A rare pictorial example of a shadow that is disconnected from the object that casts it. Charles Bentley (English, 1806–1854), after Henry Alken (English, 1785–1851), *Full Cry, from Fox Hunting*, 1828. Image credit: The Art Institute of Chicago/Art Resource, NY.

what is often seen in natural settings. For instance, they may not be properly oriented with respect to the light source (see fig. 10.32 in sec. 10.3.4), and yet they still reliably anchor the object to the ground.

10.2.2 Surface-Enhancing Shadows

Surfaces have slants and shapes that present difficulties for the painter. How can one represent the various angles of the different surfaces? We know that cast shadows can help here, although, as we shall see, some biases and taboos prevent them from being fully used, in particular at convex and concave corners (see sec. 10.3.1.2), and although (as in the Mach cards discussed in chap. 4) other pictorial cues can trump the cues provided by shadows.

Painters most frequently depicted shadows of (partly) suspended objects, shadows climbing steps (fig. 10.3), or shadows of protruding objects. We see an early example of shadows informatively interacting with surfaces in Arcangelo di Cola's *Entombment* of 1425 (fig. 10.4). Three crosses cast

Figure 10.3
The broken surface of the stairs is enhanced by the shadow of the railings. Biagio d'Antonio (d. 1516) and Jacopo del Sellaio (1441/42–1493), *Scenes from the Story of the Argonauts* (detail), ca. 1465. New York, Metropolitan Museum of Art, accession number 09.136.1.

shadows, one of which, "falling on a concave slope, is curved. These long shadows are ... unique in all of fifteenth century painting," the art historian Millard Meiss claimed.

In Carlo Crivelli's aptly named *Candle Madonna* (*Madonna della Candeletta*, fig. 10.5), we see a candle mysteriously floating in midair, possibly sticking to the plinth. The candle's shadow follows the shape of the plinth, thus assisting its visual interpretation. The shadow both fulfills an anchoring function, as it helps to disambiguate the location of the candle (without the shadow, the candle could be at any position between the plinth and the observer), and clearly indicates the shape of the plinth.

Figure 10.4

An early example of shadows interacting with surfaces: the three crosses cast shadows that tentatively reveal the slant of the hill. Arcangelo di Cola (early fifteenth century), *Entombment* (detail), 1425. Florence, private collection. Image credit: Photographic collection, Warburg Institute.

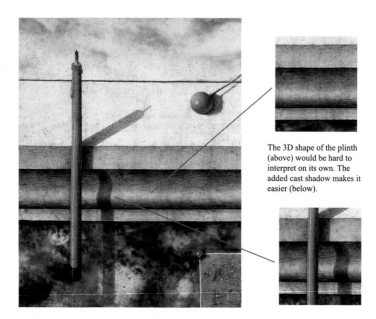

The 3D shape of the plinth (above) would be hard to interpret on its own. The added cast shadow makes it easier (below).

Figure 10.5

In and of itself, the image of the plinth (detail, *top*) is hard to interpret. The added cast shadow makes the interpretation easier (*bottom*). Carlo Crivelli (ca. 1435–1495), *Madonna della Candeletta* (detail), after 1490. Milan, Pinacoteca di Brera. Image credit: Pinacoteca di Brera.

10.2.3 Recognizable Shadows

Recognizable shadows (chap. 7) both help recognize the objects that cas
them and provide ready-made solutions to the shadow ownership problem
By reproducing some characteristic image features of the object, a recogniz
able shadow enables the viewer to recognize the object casting the shadov
based on the shadow alone, either individually (Anna, Fido) or generically
(a girl, a dog). As we saw in chapter 5, recognizable shadows can help solve
the shadow ownership problem of assigning an object to the shadow it casts.

Recognizable shadows make the objects casting them clearly recognizable: looking at the shadow suffices, provided we have a mental descriptor
for the object casting the shadow. The object itself can be either present
in the scene or absent (see the next section for the case where the object
casting the shadow is not in view). This fact indicates that although recognizable shadows may take advantage of the copycat strategy, this is not
mandatory. On the other hand, as we saw in chapter 5, a copycat shadow
will not be a recognizable shadow if the source object is itself unfamiliar.

However, a recognizable shadow may misrepresent the object that casts
it, or represent it as being an object of a different type. Contemporary artists such as Henrietta Swift or Sue Webster and Tim Noble have played with
the representational capabilities of shadows, in general stressing a conflict between the object depicted in the shadow and the object that casts
it. These works hint at the dichotomy between object and shadow. They
destroy a standard expectation about shadows: that shadows are diagnostic
of an object's shape. Actually, the artworks do show us one of the shapes of
the object, but not one that we would consider representative. The conflict
between the object's shape and its shadow in these cases voids the shadow's
diagnostic power.

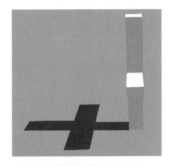

Top and middle: A copycat shadow may not be recognizable, and a recognizable shadow may not be copycat. *Bottom*: Douglas Hofstadter's "trip-let" with the initials of Gödel, Escher, and Bach plays with the strangeness of recognizable shadows. Image credit: Douglas Hofstadter.

10.2.4 Access Shadows

Gombrich singled out the case where the object casting the shadow is
absent in the scene, and he referred to them as "access" shadows. These
can be all the more important when the shadow is a recognizable one. The
example of Gérôme's *Consummatum est* (discussed in chap. 1) is one of the
few available in the art historical corpus. In natural settings, shadows are
often cast by objects outside our field of view. Nevertheless, painters in general refrain from including shadows of objects not visible in the depicted
scene, thereby cordoning off the depicted space from the surrounding
space. One counterexample is Emile Loubon's *View of Marseille* (fig. 10.6);

Figure 10.6
Emile Loubon (1809–1863), *Vue de Marseille prise des Aygalades un jour de marché*, 1853. Marseille, Musée des Beaux-Arts. Reprinted in André Gouirand, *Les peintres provençaux* (Paris: Société d'éditions littéraires et artistiques, 1901), 24. Image credit: Archive.org.

art historians interpret the dark regions at the bottom right as shadows of horsemen who are farther to the right. The shadow a photographer casts into the photograph is another example that we discussed, along with the example by Renoir (chap. 2)—cases, we may say, of "self-portrait" shadows. On top of signaling the presence of unseen characters, access shadows create an impression of a larger space than the one strictly represented in the image. Gombrich mentioned the artworks by Emanuel de Witte and William Collins as examples of the narrative power of shadows to reveal the part of the picture that is hidden from the viewer. Access shadows need not be easily recognizable, as Loubon's image testifies.

An art historical fact of the matter:

There are very few access shadows.

This paucity leads to the following intriguing fact.

10.2.5 Shadows Can Be Used to Infer the Original Properties of a Painting
The informational uses of shadows can extend beyond our understanding of the depicted scene. Shadow-based inferences can contribute to restoration

Figure 10.7
Art historian John Shearman argues that two faint shadows can be seen here, one to the left of the angel's wing, and one on which the angel sits, both cast by objects outside the painting. This suggest that the painting was originally larger than it is now. Masaccio (1401–1428), *The Virgin and Child* (detail), 1426. London, National Gallery. Image credit: Arielle Veenemans.

techniques. In the House of Augustus in Rome, postulating objects that likely cast shadows helped archaeologists put together the pieces of the frescoes that were found scattered on the floor. In Masaccio's Pisa polyptych (fig. 10.7), art historians used the shadows of nondisplayed figures to infer the size of the original panel, now incomplete. The reasoning here can only work under the assumption that it is extremely rare to have shadows cast by a figure that is not represented in the painting (access shadows). The evidence for this is purely observational: as mentioned in the previous section, very few access shadows appear in the art corpus. The art historian John Shearman argued that the Masaccio *Madonna* in the National Gallery was probably the central part of a single painting that had been broken into three. Central to Shearman's argument were two faint shadows falling from the left on two levels—cast originally, he surmised, by figures standing on the steps—which continued outside the boundary of the surviving panel.

But the conventions concerning shadows are not limited to the rarity of access shadows from objects outside the scene; they extend as well to the rule's complement: that shadows from objects in the scene should not extend outside the scene. It is as if an object and its shadow form an unbreakable molecular pictorial unit. Under this "quarantined-crowd" principle, it looks as if painters tried to create complete worlds in which, if shadows are cast, each visible shadow is complete and is the shadow of an object in the painting. The quarantined-crowd principle has two parts. The first is the access principle described earlier that discourages objects outside the frame from sending shadows inside, combined with the second principle:

In the norm, shadows do not reach outside the depicted scene.

Together, these two principles act to create the quarantined effect that was largely observed throughout the Renaissance (thus justifying Shearman's quest).

10.3   Pictorial Struggles with Shadows

What did painters do with shadows? We embark on a journey that will take us through a number of extraordinary pictorial experiments—some successful, some less so, but all interesting. We have singled out some broad categories of solutions to pictorial problems: depicted shadows having trouble negotiating obstacles in their path; shadow shapes and colors that stretch credibility; inconsistent illumination in the scene; and shadow character getting lost. We also find some taboos, that is, self-inflicted limitations on where or what to depict of a shadow.

Some of the shadow oddities in our tour are scientifically remarkable, as they suggest that painters discovered important shortcuts in representing three-dimensionality. Importantly, painters point out the visual system's peculiar tolerance for physically impossible shadows, as well as an occasional intolerance of quite possible shadows. Some other oddities simply indicate the difficulty in solving a representational problem but are interesting nonetheless. In general, most of the cases we examine involve radical simplifications of the physical aspects of the represented scene and reveal

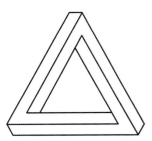

Figure 10.8
The Reutersvärd-Penrose triangle is quickly seen as being an impossible solid body.
Image credit: RC.

a (relative) tolerance in the human visual system for these simplifications. In most cases, detecting these odd physical aspects requires careful observation, which indirectly proves the thesis we are making: that some pictorial shortcuts are indeed effective. Spotting impossibilities in a picture with shadows is not like quickly noticing that something is wrong with the Reutersvärd-Penrose impossible triangle (fig. 10.8). So although we are perplexed by some impossible solid objects, we seldom notice equally implausible shadows.

In many cases, the rules of physics that apply in a real scene appear to be optional in a painting; they can be obeyed or ignored at the discretion of the artist to enhance the painting's intended effect. Some strong deviations, such as Picasso's skewed faces or the wildly colored shadows in the works of the Fauvist school (see fig. 6.31), are meant to be noticed as ingredients of the style and message of the painting—they serve communication purposes. On top of that, an alternative physics operates in many paintings, one that few of us ever notice but is just as improbable. These transgressions of standard physics—impossible shadows, impossible colors, impossible reflections or contours—often pass unnoticed by the viewer and do not interfere with the viewer's understanding of the scene. *Because we do not notice them, transgressions of physics reveal that our visual brain uses a simpler, reduced physics to understand the world.* Artists can endorse this alternative physics precisely because these particular deviations from true physics do not matter to the viewer: the artist *can take shortcuts,* presenting cues more economically and arranging surfaces and lights to suit the message of the piece rather than the requirements of the physical world. In discovering these shortcuts or strategies of image compression, artists act as research neuroscientists or as visual hackers, and we can learn a great deal from tracing their discoveries. The goal is not to expose the "slipups" of the masters but to understand the human brain. Art in this sense is a type of found science—science we can do simply by looking.

10.3.1 Shadows Meet Obstacles: How Do They Behave?

10.3.1.1 Masaccio: Carpet shadows Shadows enhance realism, but they are difficult to depict. Painters interested in shadows moved one step at a time, by trial and error, testing the limit of the tolerance of the visual brain. The strategy pays off insofar as a large tolerance of impossibility simplifies the painter's work. An intriguing example concerns the interaction of cast shadows with objects that may be placed in between the object casting the shadow and the larger intended projection screen (the ground or a wall).

In Masaccio's *Tribute* in the Brancacci Chapel in Florence (fig. 10.9), well-delineated (and in this sense quite exceptional for their time), complete shadows of the various characters appear to be cast on the ground, without

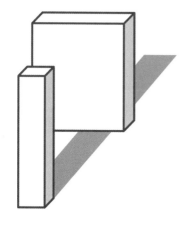

Carpet shadows pass under objects as if the objects were not present.

Figure 10.9
Masaccio (1401–1428), *Tribute*, 1425. Florence, Cappella Brancacci. Image credit: The Yorck Project.

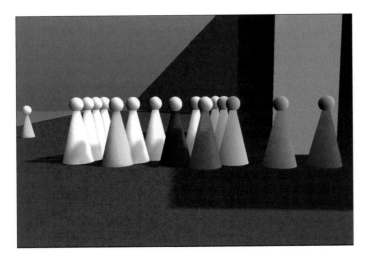

Figure 10.10
A CAD-generated model of Masaccio's *Tribute*. The shadows climb the bodies of the characters. Image credit: Meeko Kuwahara.

falling on any of the people who should intercept them. We call these "carpet shadows," as they can be walked over like carpets.

In the simulation by Meeko Kuwahara (fig. 10.10), we see what the patterns of Masaccio's shadows should have looked like. We may speculate about whether Masaccio did not have the resources for depicting the shadows correctly, or whether he chose not to depict them on bodies for aesthetic reasons. Whatever the case, the tolerance for carpet shadows is an intriguing discovery that we can usefully compare to the symmetric case of tolerance for missing cast shadows in the presence of attached shadows (chap. 4).

A powerful example of a carpet shadow appears in a painting by Konrad Witz (fig. 10.11). One possible interpretation of the image is that the shadow of an implement is cast on the floor—and spares Saint Catherine, either stopping at her dress or stopping and then reappearing behind her. According to some observers, the triangle-shaped shadow to Catherine's left is relatable to the shadow arriving from the bottom right (this would represent one of the rare access shadows found the artistic corpus). In any case, the shadow in the bottom right corner *stops mysteriously* at the saint's dress. What happened?

Figure 10.11
Saint Catherine (*right*) sits on a shadow that arrives from the bottom right and stops
at her red dress but then continues again behind her. Konrad Witz, *Sainte Madeleine
and Sainte Catherine*, ca. 1440. Strasbourg, Musée de l'Oeuvre Notre-Dame. Image
credit: Photo Musées de Strasbourg, M. Bertola.

10.3.1.2 Everybody's problem: Physical edges and chopped shadows

There is more. Carpet shadows vanish and then reappear beyond the bodies they spare. In contrast, "chopped" shadows just stop and vanish when they reach a surface where the continuation of the shadow shape and color would create pictorial difficulties for the painter. (Note that these physical implausibilities do not destroy the shadow character of either carpet or chopped shadows.)

Surface discontinuities, in particular corners where two walls meet or where the floor meets a wall, create a technical barrier to shadow representation. Already in Piero della Francesca's work, shadows get truncated at surface discontinuities (fig. 10.12).

These discontinuities typically correspond to a double possible luminance change, in either reflectance (a floor and a wall) or illumination (the two surfaces that meet are oriented differently relative to the light), or both, making it hard for the painter to compute and adjust the reflectance of the pigment for the shadow depiction. Shape recalculation is a complex

Figure 10.12
Shadows do not continue from the floor up the wall on the right. Piero della Francesca, *Polittico di Sant'Antonio* (detail), 1460–1470. Perugia, Galleria Nazionale dell'Umbria. Image credit: © Galleria Nazionale dell'Umbria.

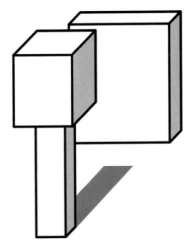

It is unlikely that a shadow ends abruptly just before a change in surface direction.

task, too: when it meets a concave or a convex corner, the shadow boundary changes direction in the image (unless the light source coincides with the viewpoint, or unless there are accidental alignments, for instance, if a change in direction in the shadow-casting rim of the object is projected precisely at the corner in the appropriate way). In what follows, we will see a number of attempts at coping with these problems—in some cases, just avoiding them; in others, coming up with arbitrary, more or less satisfactory shortcuts.

The threat of inconsistency does not appear to be a strong deterrent. In Lucas Cranach The Elder's *Der Sterbende* (fig. 10.13), the shadow of one of the two characters does, and the other does not, climb the vertical surface once it reaches it.

Figure 10.13
Not all shadows are created equal: some do not climb vertical surfaces (the shadow of the man on the left does not climb the front of the chest), but some do (the shadow of the man on the right). Lucas Cranach the Elder (1472–1553), *Der Sterbende, Epitaph des Heinrich Schmitburg* (detail), 1518. Leipzig, Museum der bildenden Künste. Image credit: Cranach Digital Archive.

There is no obvious explanation for the climbing taboo. The painter could easily diminish the cast shadow (reducing it to a small anchoring shadow near the feet, for example) so that no abrupt interruption would occur farther on where the two surfaces meet. In some cases, as the interruption coincides with a physical discontinuity, the shadow risks losing its shadow character; but this does not usually happen. On the other hand, loss of shadow character threatens shadows that are not made to turn around corners, which may explain why, while some shadows do not bend at corners as they should, others are depicted turning around corners, some correctly, some not.

In natural situations, shadows do bend at corners, if the object casting the shadow is in a suitable position (fig. 10.14). Painters have observed and depicted shadows that climb steps, bend around corners, and climb up walls (fig. 10.15).

Figure 10.14
Shadows bend around corners if the object is in the appropriate location relative to the light source and surface. However, after bending, shadows are largely deformed and generally elongated (*right*). Image credit: RC, Versilia, August 2015; Paris, April 2013.

Figure 10.15
Some represented shadows do climb steps correctly, successfully negotiating surface discontinuities. Vittore Carpaccio (1473–1526), *Presentazione della Vergine al Tempio* (detail), 1502–1505. Milan, Pinacoteca di Brera. Image credit: Pinacoteca di Brera.

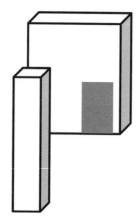

In yet other cases, the painter simply neglects the structure of the projection surface. In Girolamo Genga's *Immacolata Concezione* (fig. 10.16), a spike shadow—already a poor approximation of the object's figure—fails to adapt to the shape of the steps. Depicting a shadow correctly climbing steps does not guarantee that it is not abruptly truncated. Even Raffaello helps himself to this shortcut (fig. 10.17).

Finally, in the staircase depicted by Biagio D'Antonio (fig. 10.3; detail in fig. 10.18), we see a shadow whose *lower half* is truncated. It does climb up the wall, but how did it get there? The corner taboo seems to be confirmed in both directions: shadows get to the foot of the wall and stop there, or they climb the wall without getting there.

Figure 10.16
The shadow does not care about the steps and appears more as a transparency. Girolamo Genga (1476–1551), *Immacolata Concezione e i quattro Dottori della Chiesa* (detail), 1516–1518. Milan, Pinacoteca di Brera. Image credit: Pinacoteca di Brera.

Figure 10.17

The shadow bends appropriately over these steps but stops when it may interfere with other characters in the scene. Raffaello Sanzio (1483–1520), *Madonna and Child Enthroned with Saints* (detail), ca. 1504. New York, Metropolitan Museum of Art. Gift of J. Pierpont Morgan, cat. 16.30ab.

Figure 10.18

The man's shadow only begins on the wall, without passing along the ground to get there. Biagio d'Antonio (–1516) and Jacopo del Sellaio (1441/42–1493), *Scenes from the Story of the Argonauts* (detail), ca. 1465. New York, Metropolitan Museum of Art, accession number 09.136.1.

10.3.1.3 The Witz experiment: Shadows that bend around corners

If some shadows do not climb walls and should, others turn around corners—and should not. The problem with corners and walls is that surfaces impose constraints that are not respected by some shadows. In Konrad Witz's four large paintings for the Saint Pierre Cathedral in Geneva, we see a number of shadows systematically bending abruptly around corners. For instance, in the *Adoration of the Magi* (fig. 10.19), the shadow of the Virgin's head bends around the corner of the house. A trade-off between

Figure 10.19

Bent shadows are a recurring theme in Witz's masterpiece. The shadow of the Virgin bends around a corner. So do the shadows of a star-shaped symbol (see margin). Konrad Witz (1400–1447), *Adoration of the Magi* and *Liberation of Saint Peter* (above and in the margin), 1444. Geneva, Switzerland, Musée d'Art et d'Histoire. Image credit: Musée d'art et d'Histoire.

When in doubt, check the world: shadows that may appear odd in a painting could turn out to occur in reality. Image credit: RC, Paris, September 2018.

recognizability of the object casting the shadow and geometric accuracy has been resolved in favor of the former. Bent shadows also show up in the niche above the main doorway, where the statue of a character playing the harp is located. Another painting in the series is *The Deliverance of Saint Peter*. The general pattern is the same as in the *Adoration*. Above the doorway, a niche hosts a statue of a character holding a six-pointed star that bends around the corner of the niche. It is clear that Witz was convinced by the solution. Note that the bend is clearly signaled by a change in the direction of the terminator; the shadows are not full-blown copycats. It is thus possible that the main visual effect sought here is the enhancement of the corner (these are thus a case of surface-enhancing shadows; see sec. 10.2.2).

Witz's example is particularly telling because of his association with naturalism and careful observation of physical phenomena, in particular phenomena related to light—shadows, reflections, transparencies. The *Adoration* is depicted on the interior panel of a triptych, one of whose exterior wings, *The Miraculous Draft of Fishes*, is alleged to be one of the first representations of a real landscape: Lake Leman, with Le Môle and Mont Blanc in the background, displaying impressive details of bubbles on water, transparencies, and underwater objects.

Although Witz's experiments with bent shadows are not completely accurate, they depict possible shadows. Other artists were not so careful and produced shadows that bent impossibly around corners. For example, an impossible bending is seen on the shadow of a column in a painting by the Master of Alkmaar (fig. 10.20). Here, as earlier, we should observe that physically incorrect as they may be, shadows that bend or stop at corners still perform their surface-enhancing function: by the change in direction of their boundary, they signal the presence of a corner.

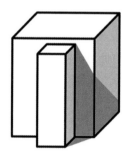

An impossible corner-turning shadow.

10.3.1.4 Truncated shadows

Edges are the main cause of shadow amputation, but shadows can also stop abruptly for no apparent reason in the middle of a perfectly continuous surface. In Piero's *Polittico di Sant'Antonio*, shadows appear to be of half columns, and not of the full columns casting them (fig. 10.21).

Truncated shadows solve in their own way the same problem addressed by carpet shadows and chopped shadows. All allow the painter to avoid struggling with the shape of shadows climbing up an item such as a human body or a wall, and with the color of the shaded area as it crosses a

Figure 10.20

The shadow of the column bends down the front of the step where the light casting the shadow cannot reach. Master of Alkmaar (Dutch painter, active ca. 1490–1510), *The Seven Works of Mercy: Tending the Thirsty* (detail), 1504. Amsterdam, Rijksmuseum. Image source: Rijksmuseum.

Figure 10.21
Truncated shadows: half columns, according to the shadow. Piero della Francesca
(1416/17–1492), *Polittico di Sant'Antonio* (detail), 1460–1470. Image credit: © Galleria
Nazionale dell'Umbria.

reflectance boundary (walls and people have in the norm different colors
from floors).

10.3.2   The Shape and Color of Shadows
What are the shape and color of shadows? Does the eye care? Did painters?

10.3.2.1   Triangle shadows   Ecologically valid in diffuse light, anchoring shadows approximate a triangular shape, in particular if the extended light source is a window open on a cloudy sky or facing north (fig. 10.22). Painters surely took good notice of the triangular shape of the umbra. Robert Campin's Merode Altarpiece (fig. 10.23) creates a dramatic setting for making the case—in an almost pedagogical way—of extended sources of

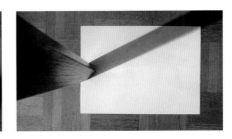

Figure 10.22

Shadow and penumbra of a vertical pole from a window open on a cloudy sky. The umbra is triangular. Note the large penumbra and the antumbra beyond the apex of the triangular umbra where the light source is visible on both sides of the table leg. The middle and right-hand images are obtained by screening part of the window with a curtain. Image credit: RC, Paris, July 2015.

Figure 10.23

Light from two openings on the left-hand wall creates double shadows, whose intersection is triangular. Robert Campin (ca. 1375–1444), *Annunciation Triptych* (Merode Altarpiece), ca. 1427–1432, and detail. New York, Metropolitan Museum of Art, accession number 56.70a-c.

light and the composition of shadows into an umbral triangular part with a residual penumbra (see detail in the margin).

Although a fastidious viewer may notice some inconsistencies in light (and a less-than-sure mastery of perspective), the painting represents a superb example of observation of shadow phenomena.

10.3.2.2 Spike shadows The theme of triangular shadow regions has many variants, and if Campin is a champion of detail, other artists resorted to less-convincing solutions up to at least the 1800s and even into Gauguin's Polynesian paintings. Sharp triangle shadows, which we may also call *spike* shadows, missing a penumbra, are a degenerate pictorial representation of triangle shadows: they have the appearance of a cast shadow from nondiffuse light, and the shape of an anchoring shadow from diffuse light. We can observe spike shadows in Roman mosaics, but also in Renaissance paintings (fig. 10.24).

Figure 10.24
Girolamo di Benvenuto (1470–1525), *Saint Catherine of Siena Exorcising a Possessed Woman* (detail), ca. October 1500. Denver Art Museum.

Figure 10.25
Inverse triangle shadows. Villa Romana del Casale, Big Game Hunt, Piazza Armerina, Sicily. Image credit: Martin Herbst.

Even more interesting, some shadows are represented as *reverse* triangle shadows (fig. 10.25), suggesting that conventionalization plays a role in shaping some shadow representations. Here neither the softness of anchoring shadows is preserved, nor the shape. We can surmise that a propositional representation has mediated the transmission of the craft ("use a triangle to depict a shadow"), and at some point in time, the orientation of the triangle reversed.

10.3.2.3 Horseshoe shadows

Horseshoe shadows are born of the attempt to represent the shadow of the pair of human legs that degenerates into a self-standing horseshoe shape (in particular, as the penumbra is absent). Horseshoe shadows have a modern, if limited, life (fig. 10.26). As with their Roman predecessors (fig. 10.27), they appear to stretch credibility. Surely we notice that something is wrong with them. The failed experiment indicates some of the limits of acceptable distortions of shadows. And yet we require little interpretive work to state that the painter's intention here was to draw shadows.

10.3.2.4 White shadows

We have seen (chap. 6) that you can tinker with shadows in many ways and still not eliminate shadow character, but one

Image credit: PC, redrawn from Paul Gauguin, *Bathers*, 1902.

Figure 10.26
Bartolomeo Cesi (1556–1629), *Amor, Anteros and Amor Lethaeus*, Palazzo Magnani, Bologna. Photo: public domain (Warburg Institute, Bodmer Archive).

thing you should never do: reverse a shadow's polarity. White shadows on a dark background not only do not exist because of physical laws but would never be perceived as shadows. The shadowed area should always have a luminance that is less than the luminance of the surface that receives them.

But what if you have already used the darkest possible color from your palette for the background surface? Mosaic artists in Lugdunum, the ancient Lyon, thought they had some margin for using lighter shadows (fig. 10.28).

Some "canonical" shadows, that is, shadows that are darker than their surroundings, are actually cast by a few objects in the same mosaic where the background is conveniently lighter.

Figure 10.27
The Magerius Mosaic, Smirat, Tunisia (3rd century). Sousse Archaeological Museum.
Image credit: Pascale Radigue.

Figure 10.28
Light shadows and colored shadows in a Roman mosaic. Circus games mosaic, second century. Lyon, Musée Gallo Romain. Image credit: Musée Gallo Romain Lyon.

Figure 10.29

"Enjoy your life." Mosaic, Antakya, Turkey, third century BCE. Image credit: Thisiscolossal.com.

A recently discovered mosaic in Antakya, Turkey, offers a dramatic example of a white shadow (fig. 10.29). This image from the third century BCE represents a skeleton who advises viewers to enjoy life. Not many shadows of skeletons exist, and the artist used a standard shape of body shadow for the sitter. The light color is once more a necessity given the dark background.

10.3.3   The Volume of Shadows

Conventionalized shadows of Roman mosaics have been variously decorated and enhanced by their makers. The decoration has at times taken the appearance of an attached shadow on the shadow itself, which in turn has conferred volume to the shadow. In figure 10.30 (which we saw previously in chap. 6), a whole catalog of ways to create "solid" shadows is displayed in a few square feet of the mosaics at the Villa Romana del Casale. The man's right leg has two shadows, one of which wraps around and passes in front of a stick that he holds.

The shadow of his left leg has its own attached shadow, which itself seems to cast a second shadow. Only one of the dog's rear paws casts a sticklike solid shadow, which is enhanced by an elaborate attached shadow. The right front paw of the same dog casts a more mundane shadow in a

Figure 10.30
Solid shadows. Hunting scene, Piazza Armerina, Villa Romana del Casale, fourth century. Image credit: Martin Herbst.

direction opposite to the other shadows in the mosaic. As we remarked in the opening section of chapter 7, one of the characters in the mosaics at the Villa Romana del Casale was even mistakenly identified as a "skier," as shadows from his feet seem take the shape of skis.

Shape, occlusion, and the presence of secondary cast and attached shadows act against our perceptual system's recognition of certain intended shadows as shadows. Yet, again, it is clear that the artist's intention was to draw shadows. It would be an interesting research question for art historians to follow the development of these modifications to shadows.

10.3.4 Incoherent, Unrelatable, or Impossible Light Sources

In some cases, it is impossible to relate shadows to any light source at all. In other cases, sources of light are inconsistent. But do we notice the inconsistencies? By 1467, artists such as Fra Carnevale had mastered consistent perspective, but not consistent lighting (fig. 10.31). The people in the foreground of Carnevale's *Birth of the Virgin* cast dark, deep shadows, but those on the piazza above and to the left do not. The alcove on the

Figure 10.31
Fra Carnevale (active 1445, d. 1484), *The Birth of the Virgin*, 1467. New York, Metropolitan Museum of Art.

Figure 10.32
Shadow chaos. Items that are parallel should cast parallel shadows. Arrows of identical color indicate the direction of shadows from legs or other items that are reasonably parallel in the scene. Luca Signorelli (1450–1523), *The Flagellation* (detail), 1480. Milan, Pinacoteca di Brera. Image credit: Web Gallery of Art. Elaboration: RC.

right is brightly lit, but the only opening in its left wall is a small door. The shadows on the alcove's right wall rise mysteriously upward, where no light source in the depicted scene could ever make them go. Nonetheless these severe inconsistencies are not evident or jarring to the human viewer.

If we assume that a single source of light is present in a scene, shadows of objects that are parallel to each other are inconsistent if they converge to a point different from the one occupied by the source. In Signorelli's *Flagellation* (fig. 10.32), many such points are present.

We can further point out double dissociations between convincing but incorrect representations and correct but unconvincing representations. Compare, for instance, the *Nativity* of Filippo Lippi in Spoleto (1466–1469) and the *Nativity* of his collaborator at Prato, Fra Diamante (1465–1470). Lippi (fig. 10.33) presents us with highly convincing shadows of poles protruding from the ruins of a building. The shadows are parallel, indicating a source of light (the sun) that is placed somewhere at the top left of the represented space. However, these shadows are impossible, as they are projected on surfaces that form an angle incompatible with the uniqueness of the light source.

The fresco is convincing, but incorrect. Fra Diamante's own elaboration on the theme (fig. 10.34), on the other hand, is correct—but unconvincing. He depicts a very similar situation, in which shadows are cast from poles protruding from two surfaces that form an angle. The shadows appear to *go in opposite directions*. Much as this can disturb the viewer, the aspect of

Figure 10.33
Inconsistent shadows by beams protruding from walls. Filippo Lippi (1406–1469), *Nativity*, 1466–1469. Spoleto, Duomo. Image credit: Comune di Spoleto.

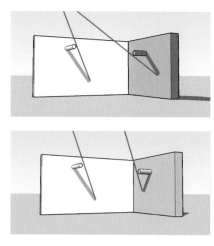

Rendering of Lippi's (*top*) and Fra Diamante's (*bottom*) respective solutions to the problem of casting shadows in a corner. The shadow rays in the top image are impossible; the bottom image is CAD generated with a virtual (sun-like) light source. Image credit: RC.

Figure 10.34
The correct solution to the Lippi problem. Fra Diamante (1430–1498), *Nativity*, 1465–1470. Paris: Louvre. Image credit: RC.

Figure 10.35
Light from above, and at the same time from below. Intarsia panel from the Gubbio
Studiolo, designed by Francesco di Giorgio Martini (1439–1501). New York, Metro-
politan Museum of Art. Image credit: RC.

the shadows is consistent with the presence of a single light source and the
relative positions of beams and walls. It looks as if Fra Diamante is correcting his master (he even places a small bird on the poles, perhaps to attract
the viewer's attention to his own performance). A comparison between the
two solutions is provided in the margin. In this case, the visual system
displays not only tolerance of the impossible but outright *preference for the
impossible*.

The intarsia from the Studiolo of the Ducal Palace in Gubbio are interesting because of some subtle inconsistencies in the shadows. In one case (fig.
10.35), the hook holding the hanging cross casts a shadow on the bottom
face of the shelf, thereby indicating that light comes from below. But at the
same time, the cross's own shadow is cast at a lower height than the cross,
thus indicating that light comes from above.

Figure 10.36
Pesellino, *Annunciation*, ca. 1455–1450. San Francisco Legion of Honor.

10.3.4.1 Lone shadows and missing shadows We should not expect consistency in shadow representation, not even within a single artwork. In artworks, it may happen that some objects cast shadows, whereas others, which should cast them as well, just do not. In some cases, a single object casts a lone shadow in an otherwise shadowless scene. As in a double dissociation pattern, we find also the dual cases of objects that do not cast shadows in scenes where other objects do (fig. 10.36). It is noteworthy that we do not seem to be particularly concerned by this shadow locality that is inconsistent with the remainder of the illumination in the scene.

10.3.4.2 Which objects are allowed to cast shadows? Opaque physical bodies block light, by definition. In Dante's *Comedy*, a cast shadow signals the presence of a body (e.g., "Purgatory," V), as opposed to a soul, which does not block light. In paintings, however, supernatural entities do not always refrain from casting shadows. For example, in the *Annunciation* by Lorenzo Lotto (fig. 10.38), the angel casts a neat shadow on the floor. But

Figure 10.37
Do ghosts block light? Do only some? Hassan Meer, *Enlightenment 1*, 2011. Image credit: Hassan Meer.

Figure 10.38
An angel can, after all, obstruct light. Lorenzo Lotto (1480–1556), *Annunciation*, ca. 1534. Recanati, Museo Civico Villa Colloredo Mels.

Figure 10.39
Souls are surprised at Dante's casting a shadow. Luca Signorelli (1445–1523), detail of
a grisaille, 1499–1502. Fresco in the Orvieto Cathedral, San Brizio Chapel.

what is an entity that it may cast a shadow? We can identify no intrinsic
ontological limitation to the type of entity that casts a shadow: it must just
be able to block light. A conceptual paradox is lurking here. To be visible,
supernatural entities must reflect light (if they do not emit it). But as a consequence they must be opaque to at least *some* light, and thus they should
cast shadows. The depiction of Dante's shadow by Signorelli in the Orvieto
Cathedral (fig. 10.39) bears witness to Dante's bodily nature and betrays
him to the (almost) shadowless souls of Purgatory.

It is, moreover, an important constraint that saints and angels, especially flying angels, need to be both visible and visually situated above the ground, in particular, to convince the viewer of their powers. Because shadows *confer a sense of presence* to objects casting them, painters must negotiate a trade-off between the visual realism provided by shadows and fidelity to the strange metaphysical properties of the objects casting the shadows.

10.3.5 Quasi Taboos

Depicted shadows dislike climbing walls; this is a pictorial disinclination, almost a taboo. Some other disinclinations have helped painters. Shadows seldom appear on human bodies, and in the corpus we examined, we found few self-shadows and few occluded shadows.

10.3.5.1 No shadows on human bodies Remember how carpet shadows do not interact adequately with human bodies. More generally, shadows are cast *on* bodies with great infrequency throughout art history. Violations of this bias are rare, the most spectacular case being the shadow of the Captain's hand in Rembrandt's *Night Watch* (fig. 10.40), and pictorially not very satisfactory. This is to be expected, in a sense. Human bodies are complex objects, and representing them is already an exceedingly difficult pictorial challenge, which includes the challenge of adequately representing patterns of light and shade on them (chiaroscuro). Cast shadows on bodies are not only particularly noisy but also extremely hard to depict. On the other hand, the narrative of the episode may place demands on its visual representation.

When the solution is relatively simple, and the enhancement obvious, shadows are easier to come by (e.g., shadows of glasses' frames on cheeks, such as those depicted by Marinus van Reymerswaele (1490–1567) in *The Tax Collectors*, now at the Hermitage in Saint Petersburg).

The taboo shadow on human body can usefully be compared with the representation of carpet shadows; but whereas carpet shadows display unsatisfactory interactions with bodies, taboo shadows avoid the confrontation altogether. Thus the implicit recommendation:

Do not place cast shadows on bodies.

10.3.5.2 Very few self-shadows Self-shadows are a particular case of cast shadows (one and the same body is part shadow caster and part receiving

Figure 10.40
If the shadow looks like a stain, it is because the ownership problem is not addressed
convincingly, and the properties of the candidate zone are not shadowy enough.
Rembrandt Harmenszoon van Rijn (1606–1669), *Militia Company of District II under
the Command of Captain Frans Banninck Cocq* (known as *The Night Watch*) (detail),
1642. Amsterdam, Rijksmuseum. Image credit: Rijksmuseum.

surface). Self-shadows too are used sparingly by painters; this lends support to the hypothesis that either the noisy character of shadows on human bodies, or the intrinsic difficulty of a satisfactory rendering (or both), makes painters wary of them. The empirical rule:

Do not represent self-shadows.

10.3.5.3 Very few occluded shadows Shadows that stem from people's feet are the norm in most paintings. In these cases, the shadow is almost always partly occluded by the foot. But apart from that case, cast shadows are seldom occluded in paintings.

It looks as if a recommendation is in place:

(Always) represent an object's shadow unoccluded.

One reason for avoiding shadow occlusion is avoiding the risk of shadow capture, as the perceptual default is that a shadow is attached to the object it is connected to in the image.

10.4 Careful Shadow Observations

10.4.1 Double Shadows

Some curious shadows make us think that if they were not depicted from memory or by calculating their appearance, they may have been the result of observation.

In *Christ before Pilate*, the Master of Cappenberg has painted both the shadow of the dog and, within it, the shadows of the dog's paws (fig. 10.41). The shadow of the left front paw is detached, indicating that the paw is lifted. But the resulting shadow image is not the shadow of a dog with paws. The paws' shadows are much darker than the shadows of the dog's body. Because the paws are closer to the ground than the body, they are less subject to the penumbral effect. (The presence of a second source of light could explain the double shadow, but that would be inconsistent with the pattern of illumination in the rest of the painting.) Another possibility is that we are in a situation similar to that of figure 6.46 (reproduced in the margin here): the contrast of the shadow of a more distant object is reduced.

Figure 10.41
Master of Cappenberg (1500–1523), *Christ before Pilate* (detail), ca. 1520. London, National Gallery. Image credit: Arielle Veenemans.

10.4.2 Virtuoso Shadows

Cappenberg's double shadow indicates that some painters, although they lacked the theoretical (geometric) background for correctly representing a shadow, could make important observations of shadow phenomena. Virtuoso shadows are based on a careful observation of a shadow phenomenon—be they copied from nature or computed offline—and are likely to be offered as a display of the artist's mastery.

A few examples, among many. We have already seen the careful rendering of double shadows by Robert Campin (fig. 10.23, detail). The opening image of this chapter is Campin's *Saint Barbara*, a wing of his *Werl Triptych*. A rich collection of double shadows and of interactions of different sources of light (note the colored shadow by the fireplace) suggests sustained visual observation and, at the same time, the blind application of practical rendering rules (e.g., the hook on one of the ceiling beams receives light from an impossibly placed source).

Figure 10.42
Each sheet of the book casts its own shadow. Marco Palmezzano (1460–1539), *Madonna con il bambino e i Santi* (detail). Milan: Pinacoteca di Brera. Image credit: Pinacoteca di Brera.

In Marco Palmezzano's *Madonna* (fig. 10.42), each page of the book casts a shadow on the next.

In *The Story of Papirius*, Domenico Beccafumi represented—indeed, dramatized—the softening at the tip of the shadows owing to the larger distance from the head to the ground, in sunlight (fig. 10.43). The image sits well with our observation that penumbras can be subject to differential foreshortening: the penumbra is only significant along the main axis of each shadow (see fig. 1.24). Beccafumi may have been the first to notice and consider the phenomenon worth representing.

10.5 Conclusions

An intriguing question should occupy art historians, and visual and cognitive scientists, as we are in the presence of a large-scale natural experiment.

Figure 10.43
Domenico Beccafumi (1484–1551), *The Story of Papirius*, mid-1520s. London, National Gallery. Image credit: Arielle Veenemans.

Why has there been no systematic depiction of shadows in the whole of world culture except for Greek and Roman artwork and then Western painting (fig. 10.44) beginning around the Renaissance?

Jacobson and Werner proposed that cast shadows are "expendable" for two different but convergent reasons. First, viewers are not sensitive to incongruities in cast shadows; second, cast shadows are difficult to tell from pigment in static displays.

The latter point can be expanded. Shadows in a picture are not subject to two important dynamic checks that are available in the real world relative to other objects in the scene. First, shadows in a picture are as flat as they are in the real world, but it is harder to check that they are just shadows and not dark patches. Second, objects in a picture are as static and "flat" as shadows, so it is harder to check that they are not just flat surfaces or decoys (in a three-dimensional environment, we could just move around them). Hence, in a picture, shadows *compete* with objects *more* than they would in the real world, and the competition is unfair. The conflict appears particularly striking when shadows are able to elicit recognition. We can call a shadow of Michelangelo's *David* a "good" shadow if it triggers the concept of Michelangelo's *David* in a viewer (see chap. 7). The shadow will visually compete with the statue for the label "David." The world of pictures is

Figure 10.44

A very early example of cast shadow used with trompe l'oeil intention. Sant'Abbondio, Como, Italy, ca. 1315–1325. Image credit: Carla Travi.

flat, and in a flat world, shadows have an advantage, precisely as they are themselves flat.

The main contribution of shadows to perception is evident precisely in pictorial perception, as they are pictorial cues. On top of the expendability of cast shadows, more mundane reasons may determine pictorial choices. Among these is that shadows are difficult to depict; or that in depicting them, the artist always risks ending up with stains, dirty spots, and visual noise. In this chapter, we discussed various pictorial functions of shadows. It is noteworthy that although shadows can serve several pictorial functions, these functions are not mutually exclusive. A given pictorial shadow can perform two or more of them simultaneously—for example, both act as an anchoring shadow and reveal features of the surface on which the shadow falls.

So imagine you want to exploit the treasure of many centuries of pictorial research to understand how painters chose to depict shadows. The path looks tortuous, but in the end it can be reconstructed as a pictorial checklist that partly recapitulates art history and the principles we have noted in passing.

0. The zero-shadow option is available, after all. You can be as cartoonish as you like: no problem in depicting no shadows, either cast or attached. Flat colors and shadowless line drawings will make for reasonably interpretable pictures.

1. However, you may want to introduce attached shadows where surfaces angled away from the light have less illumination. Attached shadows greatly enhance realism and are relatively simple to draw. And once you have the attached shadows, do not care about depicting cast shadows, as they are not easy to depict (nobody will notice; and nobody will notice if attached shadows are not matched by cast shadows). If attached shadows are predictable on a single, relatively uniform surface and it is easy to state rules of thumb for them, cast shadows may fall on complex surfaces and cross reflectance borders or illumination borders or both, and here computations are not trivial.

That said, cast shadows do greatly enhance the sense of presence of the bodies casting them. If you still want to depict cast shadows, then:

2. Keep them small. Anchoring shadows require minimal area and thus do not risk crossing reflectance or illumination discontinuities.

If you can't help it and want significantly larger shadows in your scene:

3. Keep them together with the object in a single visual unit (avoid occluded shadows, and mind the risk of shadow capture). Quarantine them. Do not introduce shadows from objects outside the frame: *There are very few access shadows.* And do not let them extend outside the frame: *In the norm, shadows do not reach outside the depicted scene. And (always) represent an object's shadow unoccluded.*

4. *Do not place cast shadows on bodies*: either on the same body (*do not represent self-shadows*) or on those of other characters in the scene; you can always make body-transparent (carpet) shadows.

5. Let shadows stop at corners: this avoids recalculating the appropriate reflectance of the pigment used for them.

6. Given the choice, prefer impossible but convincing shadows to correct but unconvincing shadows.

Finally,

7. Do not care about metaphysics: angels and other supernatural entities may, or may not, cast or receive shadows.

In the last set of recommendations, tolerance for mistaken or impossible shadows is a delicate issue and will have to be assessed on a case-by-case basis. (This is precisely where the artistic corpus becomes so immensely interesting for vision science.) It is because we do not notice them that

Figure 10.45
Like Campin, Palmezzano is a shadow virtuoso, as testified by the complex detail of the pedestal. Observe how he violates the shadow-on-body taboo (Saint John's right foot casts a shadow on his left foot), but how at the same time he falls victim to the shadow-that-stops-at-a-corner bias (the yellow robe casts a shadow that does not climb the pedestal). Marco Palmezzano (1460–1539), *Madonna con bambino benedicente in trono e i Santi Giovanni Battista e Filippo Benizzi*, ca. 1510. Cesena, Galleria dei dipinti antichi della Fondazione e della Cassa di Risparmio di Cesena. Image credit: Carlo Vannini, Reggio Emilia.

transgressions of physics reveal that our visual brain uses a simpler, reduced physics to understand the world. You can count on a lot of tolerance, but you should make sure you are not overstepping. Inverted triangle shadows, carpet shadows, horseshoe shadows, pulpy shadows with a volume of their own, and outlined shadows are not going to work that well; after all, they risk not being perceived as shadows, and they risk not accomplishing shadow functions. Tolerances may increase for shadows in motion pictures, but the realm of static pictures is more constrained.

Further Research Questions

1. Which of the "disinclinations" we have described (e.g., no shadows cast on human bodies) are true for modern and contemporary representational art? Which of the more extreme deviations that contemporary art imposes are cognitively significant?

11 Shadows in Art: A Methodological Appendix

How do we establish that a shadow is represented in a painting? And how do we ascertain that the represented shadow is cognitively interesting—for instance, that it violates a physical or a geometric constraint? In our research, we could not investigate all the intricacies of pictorial vehicles used to convey the sense that a shadow is present in the scene. Here we present a small selection from a set of problems and methodological caveats, some of which deserve further study.

11.1 Diverse Pictorial Means, Aging

First of all, there are many *means* involved in depicting shadows. These means can be as varied as to include "altering" the color of the shadow zone by painting it over covering it with tiny brushstrokes, or creating an autonomous patch of a different color. This works for paintings. Other means include hatching in drawings (in general, the medium largely affects both what can be represented and how shadows are rendered) and on intarsia, where shadows are represented by actual pieces of wood and the sfumato (shading) is represented with small wood insets. Another topic of interest is the discrepancies between preliminary versions and the final paintings. It is also arguably difficult to distinguish between original shadows and shadows added during restoration. It would be worthwhile to test whether post-Renaissance copies of Renaissance paintings add or otherwise "correct" or "improve" shadows.

11.1.1 The Gubbio Case: Lost Shadows
The intarsia panel from the Ducal Palace in Gubbio (1478–1482), created at the request of Federico da Montefeltro, based on drawings by Francesco

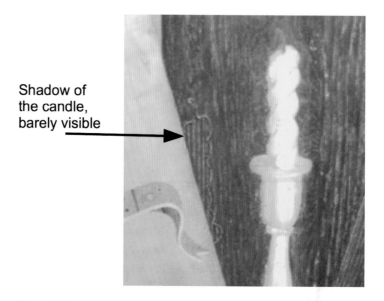

Shadow of
the candle,
barely visible

Figure 11.1
Intarsia panel from the Urbino Studiolo. New York, Metropolitan Museum of Art.
The shadow of the candle is identifiable only because its texture contrasts with that
of the adjacent wood. Image credit: RC.

di Giorgio Martini (1439–1502) and executed by Giuliano da Maiano
(1432–1490) in his Florence workshop, displays a trompe l'oeil interior with
scenes related to instruments of the liberal arts, in particular mathematics
and geometry. The elements of the panel are small wooden blocks, walnut,
beech, and rosewood, exquisitely carved to fit in a high-precision jigsaw.
Shadows abound and vividly contribute to the visual effect. In this particular medium, usually the area corresponding to a shadow is represented by a
piece of wood that is darker than an adjacent piece. The vehicle of shadow
representation is thus a solid, three-dimensional object.

Over time, some woods have darkened, so that contrast has faded, and
the shadow is visible only because the texture of the shadow block differs
from the texture of the adjacent block (fig. 11.1).

In other cases, the contrast has reversed, so that the shadow area now
has a higher luminance than the adjacent area (fig. 11.2).

11.1.2 The Sassetta Problem: Canceled Shadows and *Pentimenti*
It goes without saying that the researcher must exercise methodological
care at each point of interpreting an image. Even if we check the historical

Figure 11.2
Intarsia panel from the Urbino Studiolo. New York, Metropolitan Museum of Art. The faint diagonal bar on the middle of the panel (connecting the drum and the box) was probably meant to be a shadow of the flute, but is no longer visible. Image credit: RC.

record, we cannot always infer the artist's exact original plans, as paintings are often reshaped—by the artists themselves, by pupils, and by restorers. Shadows are potential victims of the changes. As an example, the shadow in Sassetta's *Stigmatization of Saint Francis* was considerably shortened in the process of painting (not shown here). The original shadow was revealed by using infrared imagery. The final shadow makes the saint headless; only an awkward fraction of the shadow of the right hand remains. However, we can draw some inferences from the final version (see notes for further details). Whoever was responsible for the final version was at some point confronted with the shape of Saint Francis's shadow; questions about the plausibility of what was depicted were as legitimate as they were for the first version. In what sense would the viewer accept the incongruity between the presence of the four fingers and the absence of the head? Wouldn't it have been better to drop the fingers altogether and have just a blob shadow, which would have worked anyway? The insertion of the fingers raises the threshold of accuracy and generates more expectations about the overall accuracy of the whole shadow.

11.1.3   The Reproduction Obstacle

We have taken great care to discuss only artworks that we have been able to see for ourselves. Photographs, and in particular photographs displayed on screens, can significantly alter the contrast values of the image, which are so important for assessing the presence and the quality of a shadow. And the (normally) reduced size of the reproduction can hide subtleties or create visual artifacts, in particular in digital environments.

11.2   General Interpretive Problems

On top of the issues related to the medium, we encounter difficulties in determining a gold standard of geometric and physical accuracy against which to measure the conformity or nonconformity of shadow representations. Moreover, even if a represented shadow does not fulfill some or even all the conditions of shadowness we listed in chapter 6, it can still be interpreted as a shadow by the conceptual system, which can rely on nonvisual information to make its assessment. Here are a couple of examples of such interpretive difficulties.

11.2.1 The Canaletto Problem: A Wrong Geometric Reconstruction May Hide a Truly Mistaken Geometry

Piazza San Marco in Venice has been a popular subject for painters. Canaletto painted the piazza many times. In the 1720 version (fig. 11.3), now at the Metropolitan Museum, the piazza is bathed in sunlight. The buildings of the Procuratie Nuove to the right cast their shadow on the piazza and on the tower.

Interestingly, the individuals within the building's shadow have their own dim cast shadows to the right as if from light reflected off the building to the left. The sharpness of these secondary shadows is inconsistent with the diffuse source but here we will focus on the large cast shadow from the right-hand building. Something appears to be wrong with the perspective of the piazza and the shape of the large cast shadow onto it. How can the boundary of the shadow (the one dented with the shadows of chimneys) cross the central lines of perspective of the decoration on the pavement? The fact is, the piazza has a trapezoidal shape, which hosts a decoration framed in a rectangular box (see figure 11.4). Thus because the lines of the

Figure 11.3
Giovanni Antonio Canal (Canaletto) (1697–1768), *Piazza San Marco*, ca. 1720. Oil on canvas. New York, Metropolitan Museum of Art, cat. 1988.162. Purchase, Mrs. Charles Wrightsman Gift. Image credit: Metropolitan Museum of Art.

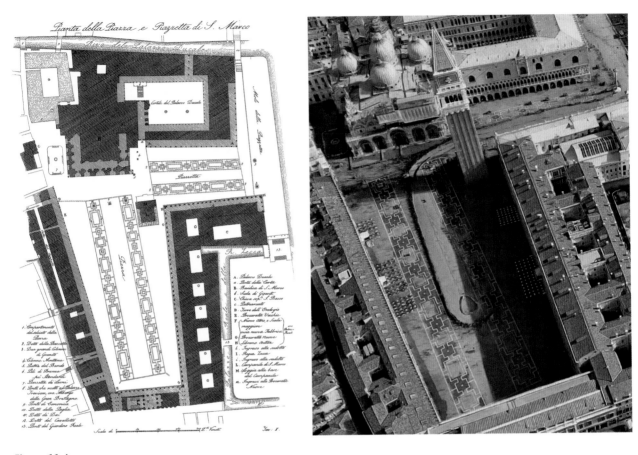

Figure 11.4

Left: Plan of Piazza San Marco, 1831. The viewpoint of Canaletto's image is at the bottom of the piazza. NNW is to the left of the image. The pavement is enclosed in a rectangular shape. *Right*: The border of the actual shadow in Piazza San Marco at a comparable time of day is parallel to the feature casting it. Image credits: Left: *Pianta della Piazza e Piazetta di San Marco*, in *La Piazza di San Marco in Venezia, descritta da Antonio Quadri e rappresentata in XVI tavole rilevate ed incise da Dionisio Moretti*. Right: Google, accessed on September 19, 2017.

floor decoration are parallel, their projections in the image converge to a point at mid-elevation in the basilica (indirectly suggesting that the viewpoint of the image is rather elevated, probably taken from the first floor of the building that closes the piazza). The projections of the bases and roofs of the Procuratie Vecchie and Nuove converge to a different vanishing point.

And yet the shadow is not correct, on different grounds. Even if we cancel out the perspective of the other buildings of the piazza, the shadow of the Procuratie Nuove ought to appear quite differently. The vanishing point for the Procuratie Nuove is situated in the second arch from the right in San Marco's facade, whereas the vanishing point for the shadow is in the tower. But if the Procuratie's roof is parallel to the floor of the piazza (as it is), that is, to the surface on which the shadow is cast, then the border of shadow of the roof should be parallel to the roof; and if the two are parallel, their projections in the image must converge, that is, share the same vanishing point (fig. 11.5). The shadow we see in this picture corresponds to a building that is taller near us and shorter near the tower. (There is actually a small epidemiology of this iconographic subject, which suggests that

Did Canaletto, or any other painter in history, use a camera obscura? The shadow test can tell: a camera obscura cannot create wrong shadows. Thus if shadows in a painting are wrong, we can at least tell that the camera obscura was not used for gathering information about *them*.

Rays of light are parallel, but because of perspective, they converge in the image. So do the beams of the ceiling. Image credit: Paolo Biagini, S. Antimo, Montalcino, Italy, January 2014.

Figure 11.5

Solid lines: parallel scene features (on the building), converging to the vanishing point. *Coarse dotted line*: Canaletto's shadow border (on the piazza) does not converge to the vanishing point. *Fine dotted line*: a plausible shadow border (depending on the time of the day). Image credit: RC.

the specific time of day was favored by painters, or that they appreciated Canaletto's display.)

This example is representative of the difficulties in interpreting the geometry of an image. You may feel certain that the shadows are wrong when they actually are not, or you may feel a shadow is accurate when it is actually incorrect. As we saw in chapter 8, forensic analysis is much concerned with these examples. In the literature, one finds spectacular cases of misinterpretation of shadows, which have led variously to claims that no humans have walked on the moon, or that evidence against Oswald in the John F. Kennedy murder case was fabricated. The classical conspiracy argument about the moon landings is that shadows in the images of the astronauts appear to converge, whereas the sun's rays are parallel. According to this logic, the shadows in the pictures should thus be cast by artificial sources of light, situated at a close distance to the astronauts. But if perspective rules the visual world, parallel lines must converge in an image (except for lines that are perpendicular to the line of sight).

11.2.2   The Van Gogh Tolerance: Non-Shadow-Looking Shadows Can Be Recognized as Shadows Anyway

Unquestionably, pictorial representations of shadows can induce recognition of shadows, even though not all conditions for shadow labeling and character are present and some conditions are counteracting.

In Van Gogh's *Painter* (fig. 11.6), the bright line around the painter's shadow acts toward destroying the shadow character (see chap. 6), most clearly in the area close to the subject's feet. At the same time, we have no doubt whatsoever that Van Gogh meant to represent a shadow—as opposed to, say, a pond of spilled oil. In a sense, this caveat is relatively harmless: the less-than-realistic rendering of Van Gogh's shadow is in line with the less-than-realistic rendering of everything else in the scene. Representational paintings are not mandatorily photorealist paintings: they strive to convey object recognition or to elicit responses to other features of the scene (e.g., emotional responses), the common denominator being that the content of the scene (trees, houses, the sky, etc.) be somewhat present to the viewer's mind by visual means only. By contrast, if an artist gave a uniformly red painting the title "Carburetor," the painting is not representational in this sense. The artist may have meant to represent a carburetor, but even if you conjure up the idea of a carburetor, you cannot do it by looking at

Figure 11.6
Vincent van Gogh (1853–1890), *Painter on His Way to Work*, 1888. Formerly at Magdeburg, Kaiser-Friedrich-Museum (probably destroyed in a fire). Image taken before 1945.

the painting only; you need to know the title. In our discussion, we focus mainly on cases that do strive toward photorealism, at the same time keeping in mind that it would be interesting to investigate less-than-photorealistic shadow character and the myriad tools that can convey it.

Once the representational purpose of the painting is ascertained, the paths to visual recognition—or to visually induced emotion arousal—are multifarious and graded. There is not a single recipe for success but a variety of more or less adequate techniques, as Gombrich pointed out. Techniques may vary; however, within a given technique, some solutions are more conducive to convincing representation to others. Photorealism is just one option.

11.3   Odd Shadows May Serve a Communication Purpose

Even the most correct shadows may appear strange. In paintings, their oddity may serve communication purposes. A communicator can make contextual elements relevant to make the receiver reconstruct an intention to convey a certain meaning. (I sit next to the closed window. She says, "It is hot in here." I understand that I have to open the window. She never explicitly said anything about opening the window; I inferred it.) In some cases we reckon that there is a communicative intention, even if we are unable to get to the intended meaning. This may be the case with some extremely curious shadows. For instance, we are inevitably bound to ask why Crivelli depicted the mysterious floating candle in his *Madonna della Candeletta* (see fig. 10.5). Is this a sort of miracle? A pictorial joke? Does it have an important meaning related to the vanity of all things? We may never be able to come up with an answer, but the interpretive question remains valid.

12 Conclusions

Visual perception has a purpose: creating representations of the environment to guide action. Its key targets are objects in space, objects that it must single out, place relative to other objects, and categorize. Its fuel is information contained in light. Light and objects are thus the core business of perception, its obsession, its benchmark. Shadows? They are in the margin. Why care about them?

Because sometimes, when we want to understand complex phenomena, it is important to look at limit cases (when you study a mathematical function, you understand a lot by considering its behavior in special cases). Shadows are precisely one of those limit cases. Perception cannot ignore them outright because they are extremely salient in their perturbation of light. Strategically, perception learned to take advantage of shadows. Looking at how perception did so, and how tortuously, shows how it works.

What did we learn along the way? We have seen that the visual system cares a lot about shadows. It knows about them even before you begin thinking about them. Its proprietary knowledge base serves the inferences that take perception from measurement to complex spatial representations of the visual scene. We also discovered that this internal knowledge is extremely rich and idiosyncratic. It does not care about the exact physical laws that govern the behavior of objects and light. For the same reason, finding out the internal rules deployed by the visual system is no minor feat. It takes a lot of ingenuity and experimenting.

We also learned that visual artists have done a great deal of the research on shadow perception. Representational art is about conveying recognition and a sense of presence of a scene: not a minor feat. Much as perception is opportunistic in relying on a simplified internal logic, so visual artists have been opportunistic in exploiting the visual shortcuts that take perception

Figure 12.1
Image credit: AP Photo/Rick Bowmer.

from measurement to inference. There is no need to reproduce a scene in all its photorealistic detail to convey recognition and a sense of presence. Thus most representational images we find in art history can be rightly considered compressed images; the history of representational art is a history of visual compression. We see small shadow stumps instead of full-blown cast shadows; attached shadows in the absence of cast shadows; a large tolerance for inconsistencies in lighting. Not all simplifications will do, of course, and the visual experiments we documented show how difficult it is to strike the balance between economy of means and richness of results. But scientists should be grateful for all the work that artists did.

Science and art are not necessarily on two opposite banks of the river of knowledge. The physicist Richard Feynman claimed that nature has a great beauty; beauty is not an accessory element of art. As the Japanese writer Jun'ichirō Tanizaki wrote in his *In Praise of Shadows*:

We find beauty not in the thing itself but in the patterns of shadows, the light and the darkness, that one thing against another creates. ... Were it not for shadows, there would be no beauty.

Glossary

Not to be considered a small encyclopedia of vision. When suitable, we refer to synonymous terms used in the literature.

Additive color mixtures   Those produced by adding lights on a single spot; the resulting luminance is higher than that of each light source.

Ambient light   Typically diffuse light existing in a scene.

Amodal completion   The perception of a whole structure when only some of its parts are visible. The hidden, or occluded, parts are represented by the visual system as present but not seen.

Anchoring shadow (Casati, 2004a) = **grounding shadow** (Jacobson & Werner, 2004)   The shadow of an object that touches a surface. If the object touches the surface, it touches its shadow (*anchoring* is preferred here because more general: an anchoring shadow can anchor the object casting it to a wall, not only to the ground).

Antumbra   The area from which the occluding body appears entirely contained within the extended light source. For example, viewers of annular eclipses of the sun are in the antumbra of the moon.

Attached shadow   "An attached shadow occurs when a surface curves away from a light source. ... Attached shadows are 1-D features of 2-D shading fields" (Hubona, Shirah, & Jennings, 2004). Also called *primary shadow*.

Bühler lines   Modifications to the appearance of a shadow by tracing other than *Hering lines*.

Carpet shadows   Depicted shadows that pass under bodies without affecting their luminance.

Cast shadow   "A cast shadow is produced when an object is interposed between a light source and a surface, blocking the illumination from reaching the surface" (Beck, 1982; quoted in Braje, Legge, & Kersten, 2000)—one definition among many. Also called a *derived shadow* (Hubona et al., 2004).

Character (e.g., shadow character).   The subjective and conscious feeling of being visually confronted with a shadow. Here opposed to *labeling*. In philosophical texts, you may find the term *quale* (plur. *qualia*). Also called *shadow identity* (Kersten, Mamassian, & Knill, 1997).

Characteristic shadow shape   A shadow shape that is easy to associate to the shape of the object casting the shadow (Sugano, Kato, & Tachibana, 2003). Also called *recognizable shadow*.

Chopped shadows   Represented shadows that just stop and vanish when they are about to hit a body.

Copycat shadow   A *recognizable* (or nonrecognizable) shadow shape profile that exactly matches the visually accessible profile of the object or objects casting the shadow (Casati, 2008).

Correspondence   See *ownership*.

Discount   "The term 'discount' is used in two senses … (i) the form of the shadow is no longer available, and (ii) this is the result of discounting the illuminant" (Rensink & Cavanagh, 2004).

Eclipse   The complete occlusion of an object by a second object.

False shadows   Artificial concoctions, typically painted over objects to suggest the presence of shadows (such as, in cosmetics, eye shadow.) Cf. *represented* and *faux* shadows.

Faux shadows   Configurations in the real world that look like shadows. Cf. *represented* and *false* shadows.

Hering lines   Reflectance lines that coincide with the profile of a shadow (Cavanagh & Kennedy, 2000).

Illumination   The light generated by one or more light sources.

Illumination edge or boundary   A division between adjacent areas in the scene due to a difference in *illumination*. Typically the line separating light from shadow; *terminators* are instances thereof. It may occasionally coincide with a *reflectance edge or boundary* and is typically accompanied by a *luminance edge*.

Isophotes   Lines of equal luminance.

Labeling (e.g., shadow labeling)   The subpersonal, unconscious identification and categorization of a part of a scene. Here opposed to *character*.

Light shadow (Elder, Trithart, Pintilie, & MacLean, 2004; Imura, Shirai, Tomonaga, Yamaguchi, & Yagi, 2008)   A "shadow" with inverted luminance, i.e., lighter than the surface on which it is cast. It has the geometry of a shadow (including consistency with the position of object, light source, and surface) and the luminance properties of a light spot. It is obviously not a shadow; the term is technical and is used for describing certain CAD-generated stimuli. At the same time, examples of represented light shadows are found in art.

Light source   The physical and geometric locus from which light emanates. It can be point-like or extended. An extended light source is typically modeled as a set of point-like sources, some of which are particularly salient for generating shadows (e.g., those on the outer boundary of the light source).

Luminance   Light reaching the eye directly from a light source or after reflection by surfaces. When a surface or a point on a surface is said to have a given luminance, that luminance corresponds to the intensity of the light reaching the eye from that surface or point.

Luminance edge or boundary   The dividing line between two adjacent regions of different luminance.

Modal completion   The perception of a structure when only some of its parts are visible. The parts that are indistinguishable from their background are seen to lie in front of it.

Mooney faces   High-contrast, two-tone images of faces, invented by the artist Giorgio Kienerk and independently rediscovered by the psychologist Craig M. Mooney.

Mutual reflection   Light reflected or diffused from one object or surface onto another. Also called *interreflection*.

Occluding boundary   Three-dimensional locus of points on object surfaces that separate visible from invisible regions; also called *rim* (Koenderink, 1984); *occluding bound* (Kennedy, 1974); and *contour generator* (Marr, 1982).

Occlusion   The partial invisibility of an object owing to the presence of a second object.

Optic array   The pencil of light rays centered on the receiving eye; alternatively, the set of environmental luminances, distributed in space, that reach the eye.

Ownership   The conscious or unconscious attribution of a shadow to its owner, i.e., the object casting it. Also called *correspondence* (Mamassian, 2004).

Penumbra   The outer boundary of a shadow, created by extended light sources. The penumbra receives light from some, but not all, of the light source.

Pictorial cues   Also called *secondary cues*. Cues to scene depth that include linear perspective, shading, shadow, occlusion, elevation, texture, texture gradients, reference frames, junctions, i.e., elements of a scene that can contribute to retrieving distance and can be represented in an image. They are opposed to *physiological* or *primary cues*, which include binocular disparity, eye convergence, and accommodation.

Projection surface   Here the surface that receives a shadow. Also termed *receiver* (Hasenfratz, Lapierre, Holzschuch, & Sillion, 2003; Sugano et al., 2003). It is one element in the standard triad constituted by object, shadow, and projection surface.

Rapid, low-level, early   The term *rapid* denotes a process carried out within the first few hundred milliseconds of processing, while *low-level* denotes processing not involving stimulus-specific knowledge; typically this is carried out in parallel across the visual field. The term *early* denotes a process that is both rapid and low-level (Rensink & Cavanagh, 2004; see also Rensink & Enns, 1998).

Recognizable shadow   A shadow that makes it easy to recognize the type or the identity of the object that casts it. It has a shape that is characteristic of the object.

Reflectance   The illumination-invariant material property of an object, its color, pigment, or paint, responsible for modulating illumination into luminance.

Reflectance edge   A line separating two regions of different reflectance. May or may not coincide with an illumination edge and in general induces a *luminance edge*.

Represented shadows   Shadows in pictures. Cf. *false* and *faux* shadows.

Reverse triangle shadows   Represented *triangle shadows* that erroneously invert the direction of the triangle.

Self shadow = intrinsic cast shadow   "Cast shadows can be intrinsic, i.e., an object casts a shadow onto itself" (Braje et al., 2000).

Shading   Shading is the variation in luminance returned to the observer's eye from an illuminated surface due to a change in the angle of surface with respect to the direction of illumination (Kingdom, Beauce, & Hunter, 2004).

Shadow   Either a *cast*, *attached*, or *self* shadow or a *shadow body*. It is one element in the standard triad constituted by object, shadow, and projection screen.

Shadow body   The volume of space where light is blocked relative to a source. It is the region of space from which an observer cannot see the source (although reflected light can reach into the region).

Shadow casting object = caster (Sharpe, 2017) **= obstructing surface** (Elder et al., 2004) **= obtruder** (Casati, 2004a) **= blocker** (Sugano et al., 2003) **= occluder** (Hasenfratz et al., 2003), **shadow caster, shadowcaster** (Rensink & Cavanagh, 2004) **= casting object** (Mamassian, Knill, & Kersten, 1998), **owner**   These terms have been used in the literature to designate the object that casts the shadow. It is one element in the standard triad constituted by object, shadow, and projection screen.

Shadow noise   Degraded information in the visual scene due to the presence of a shadow. "Shadows have long been a bane of computer vision, creating illusory objects and obscuring true object boundaries" (Fitzpatrick & Torres-Jara, 2004). "Shadow noise causes difficulties for a number of reasons. It is often quite structured, and this is not well modeled by stochastic techniques. Its effects and severity are difficult to predict: if the sun is momentarily obscured by a passing cloud ... the prevalence and effect of shadow noise can vary dramatically on time scales of less than a second" (Ollis & Stentz, 1997).

Shadow ray   A geometric construction, the line connecting a point of an object with its cast shadow. Probably first introduced by Leonardo da Vinci.

Shadow separation   Distance between the shadow-casting contour and the shadow contour (Khuu, Khambiye, & Phu, 2012).

Silhouette, outline   Projected image of the *rim* (Norman et al., 2009); "a special case of cast shadows" (Norman, Dawson, & Raines, 2000).

Spike shadows, triangle shadows   Examples of simplified represented pointed *anchoring shadows*.

Subtractive color mixtures   Those produced by filtering light from a single spot: the resulting luminance is lower than that resulting from each filter separately.

Terminator   Line dividing an attached shadow from lit-up areas of an object. Also called *generator*: "the locus of points on the surface of an object (a 3-D space curve) that projects to the shadow on a background surface" (Norman et al., 2009).

T-junction   The junction of lines in the form of a T. T-junctions are in the norm a cue to occlusion.

Triangle shadows   See *spike shadows*.

Umbra   The portion of a shadow from an extended light source that does not see any of the light source.

Visual angle   The angle subtended at the viewpoint by a dimension of an object.

X-junction   The junction of lines that cross each other in the form of an X. They are in the norm a cue to transparency and shadowness.

Notes

This section contains a general guide to the literature on shadows and perception, following the roadmap of the chapters. It selects some relevant items from the extensive reference list that follows.

Introduction

Shadows are lesser entities: Casati and Varzi (2007).

Galileo's discovery of phases of Venus as fatal blow to Ptolemaic astronomy: Drake (1957).

The history of shadow discoveries: Casati (2003).

Shadows are holes in light: Baxandall (1995).

Art director Dona Wong considers shadows as noise and discourages their use in infographics: "Don't create shadow behind bars. A bar chart is not a piece of fine art. The shadows contain no information or data": Wong (2010), p. 62.

Painters make neuroscientific discoveries: Cavanagh (2005).

Complex cultural uses of shadows: Stoichita (1997); Sharpe (2017).

The "Arnheim move": Arnheim (1974).

Chapter 1

The shadow trio: Cook (2011).

Notion of shadow body: Casati (2001).

Information in the visual scene: Gibson (1986).

Alhazen on light: Lindberg (1967, 1976).

Shadows and projections: Poncelet (1822).

Informational properties of cast shadows: Knill, Mamassian, and Kersten (1997); Bauer (1987); Da Costa Kaufmann (1993); De Rosa (1997); Allen (1999). Are shadows information or noise? Dee and Santos (2011).

Information about existence: Casati (2004b).

Massaging photographs of Soviet nomenklatura: King (1997).

Information about position: Kersten, Mamassian, and Knill (1997); Mamassian and Kersten (1996).

Antisolar point and anticrepuscular rays: apparently first described by Jesuit Father Bouvet in China in 1693, as reported by Dechevrens (1882); other reports by Hopkins (1883), as quoted by Hamblyn (2017).

The Jantar Mantar in Jaipur: Volwahsen and Jayasiṃha (2001).

Size from shadow: Casati (2016).

Distance from shadow: Kersten, Knill, Mamassian, and Bülthoff (1996); Kersten, Mamassian, and Knill (1994, 1991); Kersten et al. (1997); Mamassian and Kersten (1996).

Kagan's shadows: Casati (2014a).

The sun cannot see the shadow it casts: Goethe (1749): "... for the hideous sees he (the sun) not / As his holy eye has not/Yet alighted on shadow."

Shape from shadow: Waltz (1975, 1972); from shading: Horn (1975).

Position of the light source from shadow: Berbaum, Bever, and Chung (1983); Hagen (1976).

Heraclitus on the moon as a bowl: Dicks (1970).

Ptolemy on the concave-convex ambiguity of distant sails: Smith (1996).

Mooney faces difficult to interpret in negative contrast: Cavanagh and Leclerc (1989).

If there is a shadow, there is a surface it falls on: Casati (2004b).

Retrieving the position of light sources from shadows: Farid (2016); Mamassian and Kersten (1996).

Colored shadows and the color of light sources: Casati (2000).

Chapter 2

Shadow bodies: Casati (2001).

Pinholes: Minnaert (2013); Stoffregen (2013).

Where was Renoir standing when he painted the Pont des Saints-Pères? Stoichita (1997), p. 105; Sharpe (2017), pp. 255–256.

Chapter 3

General texts on how vision works: Marr (1982); Gregory (1998); Enns (2004); Biederman (1987); Palmer (1999); Rock (1985); Ullman (1996).

Independence of cognition and vision: Pylyshyn (1999).

On some penetrability of early visual processing: Peterson and Gibson (1991).

Face specific cells isolated in epilepsy patients: Quiroga, Kreiman, Koch, and Fried (2008).

Unconscious inference: Alhazen and Sabra (1989); Howard (1996); Helmholtz (2011).

Visual cognition and top down influences: Cavanagh (1991, 2011).

Contour and surface completion, border ownership: Kellman and Shipley (1991); Nakayama and Shimojo (1992).

Physiology of border ownership: Qiu, Sugihara, and von der Heydt (2007).

Object based completion: Tse (1999a, 1999b).

Lightness constancy, correcting for the illuminant: Adelson (2000); Gilchrist et al. (1999); Gilchrist (2006); Gilchrist and Radonjic (2010).

Objects: Spelke (1990); Feldman (2003).

Understanding visual events: Heider and Simmel (1944); Michotte (1963); Zacks and Tversky (2001).

Central conscious bulletin board: Baars (1993); Dehaene (2001).

Chapter 4

There are non-human animal precursors to the use of shadows for gaining information about the environment: Badets, Toussaint, Blandin, and Bidet-Ildei (2013); Liden and Herberholz (2008); Dally, Emery, and Clayton (2004). Chimpanzees 6–12 months old perceive pictorial depth defined by shading information: Imura and Tomonaga (2003). Depth perception based on cast shadows first appears between 5 and 7 months of age: Yonas and Granrud (2006).

Things we know about the shadow mission: e.g., cast shadows can be used to infer the shape of the casting object: Yonas, Goldsmith, and Hallstrom (1978). Shadows can be used for recognition of objects: Lovell, Gilchrist, Tolhurst, and Troscianko

(2009), and depth: Kersten et al. (1996, 1994, 1991, 1997); Mamassian and Kersten (1996).

Interference of cast shadow and recognition: Braje, Legge, and Kersten (2000); Becchio, Mari, and Castiello (2010).

Shadows are assumed to be cast from an overhead light: Mamassian and Kersten (1996); Khuu, Khambiye, and Phu (2012); chickens raised with light from below: Hershberger (1970).

Mooney-Kienerk figures: Mooney (1957); two-tone images of novel objects with shadows are difficult to recognize, whereas familiar objects, such as a Buddha, are easy: Moore and Cavanagh (1998). Recognition of two-tone structures by infants: Farzin, Rivera, and Whitney (2010); by newborns: Leo and Simion (2009).

Shadows and noise: Casati, Cavanagh, and Santos (2015).

Expendability of shadows: Jacobson and Werner (2004).

Shadow demotion: Rensink and Cavanagh (2004).

Shadow conflicts: Kersten et al. (1997).

Linking pieces of a shadow together: Casati (2012b).

Steps to the accomplishment of the shadow mission: Mamassian (2004), p. 1279.

Chapter 5

Gestalt rules: Kanizsa (1979).

The shadow ownership/correspondence problem: Mamassian (2004). Mamassian uses the term *correspondence*; we prefer the more specific *ownership*.

Gestalt rules for parsing the visual scene: Palmer (1999).

Poor understanding of eclipses: Barnett and Morran (2002); Baxter (1989); Broadstock (1992); Schoon (1993).

Shadow capture: Casati (2012b).

Copycat shadows: Casati (2008).

On Giovanni di Paolo: Sabler (2004); Meiss (1968).

On shadow occlusion: Santos, Casati, Dee, Schultz, and Bhatt (2015).

On Belbello da Pavia: Ludovici (1954).

On Magritte: Gablik (1970).

On Vasari: Stoichita (1997).

The perceptual preferences of the herring gull chick: Staddon (1975); Perdeck and Tinbergen (1951).

Chapter 6

On the perception of cast shadows: Mamassian, Knill, and Kersten (1998).

On Dalí: Sabler (2004).

Logvinenko's figure: Logvinenko (1999).

Physical properties of shadows: Cavanagh and Leclerc (1989). Only the darkness property is mandatory: Cavanagh and Leclerc (1989). X-junctions: Adelson (2000).

A region can be taken as a shadow even if its shape is incorrect: Mamassian (2004).

Shadow must be darker at their edge: Cavanagh and Leclerc (1989).

Mooney faces are disrupted by contrast reversal: Cavanagh and Leclerc (1989).

Shadows must look flat: Cavanagh (2005).

Shadows cannot be opaque or object-like: Cavanagh (2005).

The Leonardo effect: do not outline a shadow, or it will look wooden: da Vinci (1923).

Hering lines: Hering (1878, 1964).

Bühler lines: Bühler (1922).

Shadows may overcome contradictions from depth or motion: Cavanagh and Leclerc (1989).

Color change more likely at material than shadow border: Rubin and Richards (1982); Mullen and Kingdom (1991); Tappen, Freeman, and Adelson (2005); Switkes, Bradley, and De Valois (1988); Cavanagh (1991); Jacobson and Werner (2004); Kingdom, Beauce, and Hunter (2004).

Nevertheless, color does not matter for human labeling of shadows: Cavanagh and Leclerc (1989); although see Kingdom, Beauce, and Hunter (2004), for a different opinion.

Shape match does not matter: Mamassian (2004); Sattler, Sarlette, Mücken, and Klein (2005).

Conflicts of illumination direction: Ostrovsky, Sinha, and Cavanagh (2005). Assumptions about light position: Mamassian and Goutcher (2001); Morgenstern et al. (2011); Tarr, Kersten, and Bülthoff (1998); Sun and Perona (1998); Xia, Pont, and Heynderickx (2016).

Cast shadows have straight boundaries: Logvinenko, Adelson, Ross, and Somers (2005).

Shadows are fuzzy and straight, paint is sharp and crooked: Adelson and Somers (2000). Mountain shadows tend to be triangular-shaped: Livingston and Lynch (1979).

Processing of cast shadows and of attached shadows are independent: Casati (2014c).

Discounting the illuminant, Gelb experiments: Gilchrist (2006); Winkler, Spillmann, Werner, and Webster (2015).

Shadow perception can be manipulated in ecological settings: Casati (2007).

Chapter 7

Equivalent lighting model where an estimate of the direction and intensity of light is available throughout the scene: Boyaci et al. (2006); Gilchrist (2018).

On cognitive impenetrability: Concepts and the lexicon: Talmy (1983); specifically on shadows: Casati (2017).

Figures (even immaterial ones) are expected to cast shadows: Casati and Sørensen (2012); color spreading generates figures: Bressan, Mingolla, Spillmann, and Watanabe (1997).

Shadows of shadows: Casati (2012b).

Umbra and penumbra: Santos, Casati, Dee, Schultz, and Bhatt (2015).

A visible boundary is required for shadow perception: Casati (2000).

Shadow bodies: Casati (2001).

Light bodies: Minnaert (2013).

Conceptual errors in relating colored shadows to colored light sources: Casati (2000, 2017).

Filter shadows: Sørensen (2007).

The Yale problem: Casati (2001); Todes and Daniels (1975).

Toddlers may treat cast shadows as objects: Van de Walle, Rubenstein, and Spelke (1998).

Shadows in myths and narratives: Frazer (1922). Narratives must be handled with some care: Casati (2000).

Shadows that enable the recognition of the object that casts them are used in art: Sharpe (2017); Stoichita (1997).

The coarser item is perceived as a shadow: Roy Sørensen (personal communication).

Proper and actual domain of visual recognition: Sperber and Hirschfeld (2004).

Recognition based on canonical views of objects: Tarr and Bülthoff (1998).

Crackdows (can be seen as either cracks or shadows): Segal (1989).

Shadows get the appearance of real objects: Casati (2007).

The metaphors of the shadow in Plato: Todes (1975); Todes and Daniels (1975).

Analogs of shadows in glaciology: Casati (2001); and acoustics: Dufour (2011). Metaphoric uses of shadows: Tversky (2017); Frazer (1922).

General properties of the conceptual system: Barsalou (2012); Jackendoff (2010); Margolis and Laurence (1999); Prinz (2002). Cognitive illusions are like perceptual illusions: Kahneman and Tversky (1996).

Chapter 8

Human vision's performance in detecting illumination inconsistencies: Lopez-Moreno et al. (2010).

Pied-de-vent: We are indebted to Luc Baronian for this piece of information.

On pinholes: Minnaert (2013), originally Minnaert (1954); Stoffregen (2013).

Actress can control her movement by looking at her shadow: Bonfiglioli, Pavani, and Castiello (2004).

About the forensics of images: Farid (2016); Johnson and Farid (2007a, 2007b); Farid and Bravo (2010); Kee, O'Brien and Farid (2013).

Chapter 9

Conceptual uncertainties about mirrors: Casati (2012a); Chalmers (2015); Mac Cumhaill (2011).

Highlights and perceptual problems of concave and convex mirrors: Cavanagh (2005).

The ecology of mirrors: Cavanagh, Chao, and Wang (2008); Fleming, Dror, and Adelson (2003); Miller (1998).

Objects in rear mirrors: Hecht, Bertamini, and Gamer (2005).

The Venus effect: Bertamini, Latto, and Spooner (2003); Croucher, Bertamini, and Hecht (2002).

People have little idea of what to expect in mirrors: Bertamini (2014); Bertamini, Latto, et al. (2003); Bertamini, Spooner, and Hecht (2003); Hecht et al. (2005).

Learning to use a mirror: Hecht et al. (2005).

The Gombrich effect (tracing one's head on the surface of a mirror): Gombrich (1960), p. 5.

Factual vs. perspectival content of perception and of mirrors: Noë (2003).

Properties of mirrors vs mirrorness: Cavanagh et al. (2008). Mirror agnosia: Ramachandran, Altschuler, and Hillyer (1997); Binkofski, Buccino, Dohle, Seitz, and Freund (1999).

Role of symmetry in mirrorness: Cavanagh et al. (2008).

Understanding of mirrors in early childhood: Suddendorf and Whiten (2001); Sayim and Cavanagh (2011).

Japanese woodcut art and perspective: Hagen (1986); and shadow depiction: Toyama and Naito (2007).

Light (and shadows) cross the mirror boundary: Gardner (2005), p. 152; Gregory (1998).

Perceiving transparent materials: Kawabe, Maruya, and Nishida (2015).

Perceptual vs. physical transparency: Metelli (1974). Computational rules for perceived transparency: Adelson and Anandan (1990); Anderson (1997); Beck, Prazdny, and Ivry (1984); Gerbino, Stultiens, Troost, and de Weert (1990); Metelli (1985, 1974).

Earlier studies of transparency: Helmholtz (2011); Koffka (1935).

Importance of X-junctions for perceived transparency: Adelson and Anandan (1990); Anderson (1997); Beck et al. (1984); Brill (1984); Delogu, Fedorov, Belardinelli, and van Leeuwen (2010); Gerbino et al. (1990); Kersten et al. (1991); Koenderink, van Doorn, Pont, and Wijntjes (2010); Koenderink, van Doorn, Pont, and Richards (2008); Masin (2006); Nakayama, Shimojo, and Ramachandran (1990); Plummer and Ramachandran (1993).

Ordinal luminance relations: Adelson and Anandan (1990); Beck and Ivry (1988).

Luminance magnitude reduce ambiguities in transparency: Anderson (1997, 2003).

Explicit X-junctions not necessary for the perception of transparency: Watanabe and Cavanagh (1996).

Color constraints for transparency: D'Zmura, Colantoni, Knoblauch, and Laget (1997); Faul and Ekroll (2002); Fulvio, Singh, and Maloney (2006).

Laciness (superposition of textures): Watanabe and Cavanagh (1996), with no need to follow luminance constraints: Sayim and Cavanagh (2011).

Which came first, transparency perception or shadow perception: Stoner (1999); Stoner, Albright, and Ramachandran (1990).

Shadows perceived as transparent film: Arnheim (1974).

Blur as a signature of shadows and light: Bressan et al. (1997).

Highlights contribute to transparency: Watanabe and Cavanagh (1996).

Chapter 10

The Gombrich program: Gombrich (1960).

Shadows with a narrative role and personality: Sharpe (2017); Stoichita (1997).

On Arcangelo di Cola: Meiss (1968).

Access shadows: Gombrich (1995); on Loubon: Gouirand (1901).

Reconstructing frescoes from Augustus's house thanks to shadows: Casati (2000); Carettoni (1983). Reconstructing Masaccio's Pisa Polypthich thanks to shadows: Gordon (2003); Shearman (1966).

The quarantined crowd principle for depicting shadows of groups of people: Casati (2014b).

An alternative physics used by the brain: Cavanagh (2005).

Ancient pictorial competitions: Kris and Kurz (1934).

Masaccio's carpet shadows: Casati (2000); Freedman (1990); Kretzenbacher (1961).

Witz's carpet shadow: Sharpe (2017).

On Konrad Witz: Elsig and Menz (2013); Mougenel (2011).

The paradox of supernatural entities casting shadows, including Dante's *Comedy*: Casati (2000, 2014b).

Shadows mistaken for real objects at Villa del Casale mosaics: Capizzi, Galati, and Ascani (1989). Epidemiology of ancient shadow representations: Pensabene and Gallocchio (2011).

Incoherent light sources in Fra'Carnevale: Cavanagh (2005); incoherent sources in Filippo Lippi: Casati (2000, 2003).

Early pre-Renaissance shadows in S. Abbondio, Italy: Travi (2011).

Cast shadows are expendable in art: Jacobson and Werner (2004).

Chapter 11

Methodological issues in the study of cast shadows in art: Casati (2004a).

On the aging of art: Taylor (2015).

An example of later copy of an original: Perugino (1445–1523), *Vision of St. Bernard* (ca. 1491–94), Munich, Alte Pinakothek, as copied by Felice Ficherelli, Florence, Church of St. Spirit, 1655–1656.

On Sassetta's shadow cancellation: Toyama (2009). Gordon (2003) observes that there is no evidence of repainting by restorers (we thank Paul Taylor for pointing this out to us).

On Canaletto's shadows of people in the Piazza: Wijntjes and de Ridder (2014).

Art historian Paul Taylor observes (personal communication, Nov. 2017) another peculiarity of the perspective: "The orthogonals of the decoration on the piazza converge to a higher horizon than the orthogonals of the buildings. Canaletto must have painted the piazza from a different vantage point to the buildings; he moved up a storey. That explains why the shadow is inconsistent with the buildings. It's consistent with the buildings as viewed from a higher vantage point."

Independent evidence that Canaletto did not use a *camera obscura*: Royal Collection Trust (2017).

Image fabrication: Farid (2016).

Imagination in pictorial perception: Walton (1990). The recognition toolbox in depiction is very rich: Gombrich (1960).

Nonpictorial, narrative functions of shadows lie outside the boundaries of the present book. They are discussed by Sharpe (2017).

References

Adelson, E. H. (2000). Lightness perception and lightness illusions. In M. Gazzaniga (Ed.), *The new cognitive neurosciences* (pp. 339–351). Cambridge, MA: MIT Press.

Adelson, E. H., & Anandan, P. (1990). Ordinal characteristics of transparency. In *Proceedings of the AAAI Workshop on Qualitative Vision*. http://citeseer.ist.psu.edu/viewdoc/summary?doi=10.1.1.69.4477.

Adelson, E. H., & Somers, D. (2000). Shadows are fuzzy and straight; paint is sharp and crooked. *Perception, 29* (ECVP Abstract Supplement), 46.

Alhazen & Sabra, A. I. (1989). *The optics of Ibn al-Haytham, Books I–III: On direct vision.* London: Warburg Institute, University of London.

Allen, B. P. (1999). Shadows as sources of cues for distance of shadow-casting objects. *Perceptual and Motor Skills, 89*, 571–584. doi:10.2466/pms.1999.89.2.571.

Anderson, B. L. (1997). A theory of illusory lightness and transparency in monocular and binocular images: The role of contour junctions. *Perception*, 26(4), 419–453.

Anderson, B. L. (2003). The role of occlusion in the perception of depth, lightness, and opacity. *Psychological Review*, 110(4), 785–801. doi:10.1037/0033-295X.110.4.785.

Arnheim, R. (1974). *Art and visual perception: A psychology of the creative eye.* Berkeley: University of California Press.

Baars, B. J. (1993). *A cognitive theory of consciousness* (repr. ed.). Cambridge: Cambridge University Press.

Badets, A., Toussaint, L., Blandin, Y., & Bidet-Ildei, C. (2013). Interference effect of body shadow in action control. *Perception*, 42(8), 873–883. doi:10.1068/p7502.

Barnett, M., & Morran, J. (2002). Addressing children's alternative frameworks of the Moon's phases and eclipses. *International Journal of Science Education*, 24(8), 859–879.

Barsalou, L. W. (2012). The human conceptual system. In M. Spivey, K. McRae, & M. Joanisse (Eds.), *The Cambridge handbook of psycholinguistics* (pp. 239–258). New York: Cambridge University Press.

Bauer, G. (1987). Experimental shadow casting and the early history of perspective. *Art Bulletin*, 69(2), 211–219. doi:10.2307/3051018.

Baxandall, M. (1995). *Shadows and enlightenment.* New Haven, CT: Yale University Press.

Baxter, J. (1989). Children's understanding of familiar astronomical events. *International Journal of Science Education*, 11(5), 502–513.

Becchio, C., Mari, M., & Castiello, U. (2010). Perception of shadows in children with autism spectrum disorders. *PLoS One*, 5(5), e10582.

Beck, J. (Ed.). (1982). *Organization and representation in perception.* Hillsdale, NJ: Erlbaum.

Beck, J., & Ivry, R. (1988). On the role of figural organization in perceptual transparency. *Perception and Psychophysics*, 44(6), 585–594.

Beck, J., Prazdny, K., & Ivry, R. (1984). The perception of transparency with achromatic colors. *Perception and Psychophysics*, 35(5), 407–422.

Berbaum, K., Bever, T., & Chung, C. S. (1983). Light source position in the perception of object shape. *Perception*, 12(4), 411–416. doi:10.1068/p120411.

Bertamini, M. (2014). Understanding what is visible in a mirror or through a window before and after updating the position of an object. *Frontiers in Human Neuroscience*, 8. doi:10.3389/fnhum.2014.00476.

Bertamini, M., Latto, R., & Spooner, A. (2003). The Venus effect: People's understanding of mirror reflections in paintings. *Perception*, 32(5), 593–599.

Bertamini, M., Spooner, A., & Hecht, H. (2003). Naive optics: Predicting and perceiving reflections in mirrors. *Journal of Experimental Psychology: Human Perception and Performance*, 29(5), 982–1002.

Biederman, I. (1987). Recognition-by-components: A theory of human image understanding. *Psychological Review*, 94(2), 115–147.

Binkofski, F., Buccino, G., Dohle, C., Seitz, R. J., & Freund, H.-J. (1999). Mirror agnosia and mirror ataxia constitute different parietal lobe disorders. *Annals of Neurology*, 46(1), 51–61.

Bonfiglioli, C., Pavani, F., & Castiello, U. (2004). Differential effects of cast shadows on perception and action. *Perception*, 33(11), 1291–1304. doi:10.1068/p5325.

Boyaci, H., Doerschner, K., & Maloney, L. T. (2006). Cues to an equivalent lighting model. *Journal of Vision*, 6(2), 2. doi:10.1167/6.2.2.

Braje, W. L., Legge, G. E., & Kersten, D. (2000). Invariant recognition of natural objects in the presence of shadows. *Perception*, 29(4), 383–398. doi:10.1068/p3051.

Bressan, P., Mingolla, E., Spillmann, L., & Watanabe, T. (1997). Neon color spreading: A review. *Perception*, 26(11), 1353–1366.

Brill, M. H. (1984). Physical and informational constraints on the perception of transparency and translucency. *Computer Vision Graphics and Image Processing*, 28(3), 356–362. doi:10.1016/S0734-189X(84)80013-4.

Broadstock, M. J. (1992). Elementary students' alternative conceptions about Earth systems phenomena in Taiwan, Republic of China. Ohio State University. http://istar.openrepository.com/istar/handle/11290/608098.

Bühler, K. (1922). *Die Erscheinungsweisen der Farben*. Jena: Gustav Fischer.

Capizzi, C., Galati, F., & Ascani, A. (1989). *Piazza Armerina: the mosaics and Morgantina*. Bologna, Italy: Italcards.

Carettoni, G. (1983). La decorazione pittorica della casa di Augusto sul Palatino. *Bollettino Dell Stituto Archeologico Germanico. Sezione Romana*, 90, 373–419.

Casati, R. (2000). *La scoperta dell'ombra: Da Platone a Galileo; La storia di un enigma che ha affascinato le grandi menti dell'umanità*. Milano: Mondadori. Translated as *The shadow club* (Abigail Asher, Trans.). New York: Knopf, 2003. Originally reprinted as *Shadows* (New York: Vintage).

Casati, R. (2001). The structure of shadows. In A. Frank, J. Raper, & J. P. Cheylan (Eds.), *Life and motion of socio-economic units* (pp. 99–109). London: Taylor & Francis.

Casati, R. (2004a). Methodological issues in the study of the depiction of cast shadows: A case study in the relationships between art and cognition. *Journal of Aesthetics and Art Criticism*, 62(2), 163–174.

Casati, R. (2004b). The shadow knows: A primer on the informational structure of cast shadows. *Perception*, 33(11), 1385–1396.

Casati, R. (2007). How I managed to hide my shadow. *Perception*, 36(12), 1849–1851.

Casati, R. (2008). The copycat solution to the shadow correspondence problem. *Perception*, 37(4), 495–503.

Casati, R. (2012a). Mirrors, illusions and epistemic innocence. In C. Calabi & P. Spinicci (Eds.), *Perceptual illusions* (pp. 192–201). London: Palgrave.

Casati, R. (2012b). Some varieties of shadow illusions: Split shadows, occluded shadows, stolen shadows, and shadows of shadows. *Perception*, 41(3), 357–360.

Casati, R. (2014a). Incident light. In *Larry Kagan: Lying Shadows* (pp. 9–12). Glenn Falls: The Hyde Collection.

Casati, R. (2014b). Looking at images and reasoning about their content: The case of shadow depiction. In P. Taylor (Ed.), *Gombrich: Meditations on a heritage* (pp. 139–153). London: Paul Holberton. http://italianacademy.columbia.edu/sites/default/files/papers/casati_10-10.pdf.

Casati, R. (2014c). Shadow-related concavity–convexity inversions reveal a very basic tolerance for impossible shadows. *Perception*, 43(4), 351–352. doi:10.1068/p7727.

Casati, R. (2016). Size from shadow: Some informational paths less traveled. *Ecological Psychology*, 28(1), 56–63.

Casati, R. (2017). Shadows, objects and the lexicon. In T. Crowther & C. M. Cumhaill (Eds.), *Perceptual ephemera* (pp. 1–12). Oxford University Press.

Casati, R., Cavanagh, P., & Santos, P. E. (2015). The message in the shadow: Noise or knowledge? (Dagstuhl Seminar 15192). *Dagstuhl Reports*, 5(5), 24-42. doi: 10.4230/DagRep.5.5.24.

Casati, R., & Sørensen, R. (2012). Non-physical visual objects generated by colour spreading are expected to cast shadows. *Perception*, 41(10), 1275–1276.

Casati, R., & Varzi, A. C. (Eds.). (2007). Lesser kinds [Special issue]. *Monist, 90*(3).

Cavanagh, P. (1991). What's up in top-down processing. In A. Gorea (Ed.), *Representations of vision: Trends and tacit assumptions in vision research* (pp. 295–304). Cambridge: Cambridge University Press.

Cavanagh, P. (2005). The artist as neuroscientist. *Nature*, 434(7031), 301–307. doi:10.1038/434301a.

Cavanagh, P. (2011). Visual cognition. *Vision Research*, 51(13), 1538–1551.

Cavanagh, P., Chao, J., & Wang, D. (2008). Reflections in art. *Spatial Vision*, 21(3), 261–270.

Cavanagh, P., & Kennedy, J. M. (2000). Close encounters: Details veto depth from shadows. *Science*, 287(5462), 2423–2425.

Cavanagh, P., & Leclerc, Y. G. (1989). Shape from shadows. *Journal of Experimental Psychology: Human Perception and Performance*, 15(1), 3–27.

Chalmers, D. (2015). Spatial illusions: From mirrors to virtual reality. Presented at the Jean Nicod Lectures, École normale supérieure, Paris. https://www.youtube.com/watch?v=RKRX6CZLZOw.

Cook, N. D. (2011). *Harmony, perspective, and triadic cognition*. Cambridge: Cambridge University Press.

Croucher, C. J., Bertamini, M., & Hecht, H. (2002). Naive optics: Understanding the geometry of mirror reflections. *Journal of Experimental Psychology: Human Perception and Performance*, 28(3), 546–562.

Da Costa Kaufmann, T. (1993). The perspective of shadows: The history of the theory of shadow projection. In *The mastery of nature: Aspects of art, science, and humanism in the Renaissance* (pp. 49–78). Princeton, NJ: Princeton University Press.

da Vinci, L. (1923). *Leonardo da Vinci's Note-Books*. Edward McCurdy (Ed.). New York: Empire State Book Company.

da Vinci, L. (1939). *The literary works of Leonardo da Vinci*. London: Oxford University Press.

Dally, J. M., Emery, N. J., & Clayton, N. S. (2004). Cache protection strategies by western scrub-jays (*Aphelocoma californica*): Hiding food in the shade. *Proceedings of the Royal Society B: Biological Sciences*, 271(Suppl. 6), S387–S390. https://doi.org/10.1098/rsbl.2004.0190.

De Rosa, A. (1997). *Geometrie dell'ombra: Storia e simbolismo della teoria delle ombre*. Milano: Cittàstudi.

Dechevrens, M. (1882). A curious halo. *Nature*, 27(680), 30–31. doi:10.1038/027030b0.

Dee, H. M., & Santos, P. E. (2011). The perception and content of cast shadows: An interdisciplinary review. *Spatial Cognition & Computation*, 11, 226–253. doi.org/10.10 80/13875868.2011.565396.

Dehaene, S. (2001). Towards a cognitive neuroscience of consciousness: Basic evidence and a workspace framework. *Cognition*, 79(1–2), 1–37. doi:10.1016/S0010-0277(00)00123-2.

Delogu, F., Fedorov, G., Belardinelli, M. O., & van Leeuwen, C. (2010). Perceptual preferences in depth stratification of transparent layers: Photometric and non-photometric factors. *Journal of Vision*, 10(2), 19.1–13. https://doi.org/ 10.1167/10.2.19.

Dicks, D. R. (1970). *Early Greek astronomy to Aristotle*. London: Thames and Hudson.

Drake, S. (1984). Galileo, Kepler, and phases of Venus. *Journal for the History of Astronomy*, 15(3), 198–208. doi:10.1177/002182868401500304.

Dufour, F. (2011). Acoustic shadows: An auditory exploration of the sense of space. *SoundEffects: An Interdisciplinary Journal of Sound and Sound Experience*, 1(1), 82–97.

D'Zmura, M., Colantoni, P., Knoblauch, K., & Laget, B. (1997). Color transparency. *Perception*, 26(4), 471–492. doi:10.1068/p260471.

Elder, J. H., Trithart, S., Pintilie, G., & MacLean, D. (2004). Rapid processing of cast and attached shadows. *Perception*, 33(11), 1319–1338. doi:10.1068/p5323.

Elsig, F., & Menz, C. (2013). *Konrad Witz: Le maître-autel de la cathédrale de Genève.* Geneva: Slatkine.

Enns, J. T. (2004). *The thinking eye, the seeing brain: Explorations in visual cognition.* New York: W. W. Norton.

Farid, H. (2016). *Photo forensics.* Cambridge, MA: MIT Press.

Farid, H., & Bravo, M. J. (2010). Image forensic analyses that elude the human visual system. *Proceedings of SPIE, 7541,* 1–10. doi:10.1117/12.837788.

Farzin, F., Rivera, Susan M., & Whitney, D. (2010). Spatial resolution of conscious visual perception in infants. *Psychological Science, 21*(10), 1502–1509. https://doi.org/10.1177/0956797610382787.

Faul, F., & Ekroll, V. (2002). Psychophysical model of chromatic perceptual transparency based on subtractive color mixture. *Journal of the Optical Society of America A: Optics, Image Science, and Vision,* 19(6), 1084–1095.

Feldman, J. (2003). What is a visual object? *Trends in Cognitive Sciences,* 7(6), 252–256.

Fitzpatrick, P. M., & Torres-Jara, E. R. (2004). The power of the dark side: Using cast shadows for visually-guided touching. In *4th IEEE/RAS International Conference on Humanoid Robots, 2004* (Vol. 1, pp. 437–449). IEEE. http://ieeexplore.ieee.org/abstract/document/1442136.

Fleming, R. W., Dror, R. O., & Adelson, E. H. (2003). Real-world illumination and the perception of surface reflectance properties. *Journal of Vision,* 3(5), 347–368. doi:10.1167/3.5.3.

Frazer, J. G. (1922). *The golden bough.* London: Macmillan.

Freedman, L. (1990). Masaccio's *St. Peter Healing with His Shadow*: A study in its iconography. *Notizie da Palazzo Albani, 19,* 13–31.

Fulvio, J. M., Singh, M., & Maloney, L. T. (2006). Combining achromatic and chromatic cues to transparency. *Journal of Vision,* 6(8), 760–776. doi:10.1167/6.8.1.

Gablik, S. (1970). *Magritte.* Greenwich, CT: New York Graphic Society.

Gardner, M. (2005). *The new ambidextrous universe: Symmetry and asymmetry from mirror reflections to superstrings* (3rd rev. ed.). Mineola, NY: Dover.

Gerbino, W., Stultiens, C. I., Troost, J. M., & de Weert, C. M. (1990). Transparent layer constancy. *Journal of Experimental Psychology: Human Perception and Performance,* 16(1), 3–20.

Gibson, J. J. (1986). *The ecological approach to visual perception.* Hillsdale, NJ: Erlbaum.

Gilchrist, A. L. (2006). *Seeing black and white*. Oxford: Oxford University Press.

Gilchrist, A. (2018). To compute lightness, illumination is not estimated, it is held constant. *Journal of Experimental Psychology: Human Perception and Performance, 44*(8), 1258–1267.

Gilchrist, A., & Jacobsen, A. (1984). Perception of lightness and illumination in a world of one reflectance. *Perception, 13*(1), 5–19.

Gilchrist, A., Kossyfidis, C., Bonato, F., Agostini, T., Cataliotti, J., Li, X., et al. (1999). An anchoring theory of lightness perception. *Psychological Review,* 106(4), 795–834.

Gilchrist, A. L., & Radonjic, A. (2010). Functional frameworks of illumination revealed by probe disk technique. *Journal of Vision,* 10(5), 6. doi:10.1167/10.5.6.

Goethe, J. W. von. (1749). *Faust, Part II, Act III*.

Gombrich, E. H. (1960). *Art and illusion*. London: Phaidon.

Gombrich, E. H. (1995). *Shadows: The depiction of cast shadows in Western art*. New Haven, CT: Yale University Press.

Gordon, D. (2003). *The fifteenth century Italian paintings* (Vol. 1). London: National Gallery Publications.

Gouirand, A. (1901). *Les peintres provençaux*. Paris: Société d'éditions littéraires et artistiques.

Gregory, R. L. (1966). *Eye and brain: The psychology of seeing*. Princeton: Princeton University Press.

Gregory, R. L. (1998). *Mirrors in mind*. London: Penguin.

Hagen, M. A. (1976). The development of sensitivity to cast and attached shadows in pictures as information for the direction of the source of illumination. *Perception and Psychophysics,* 20(1), 25–28.

Hagen, M. A. (1986). *Varieties of realism: Geometries of representational art*. Cambridge: Cambridge University Press.

Hamblyn, R. (2017). *Clouds*. http://www.overdrive.com/media/3380903/clouds.

Hasenfratz, J. M., Lapierre, M., Holzschuch, N., & Sillion, F. (2003). A survey of real-time soft shadows algorithms. *Computer Graphics Forum,* 22(4), 753–774. doi: 10.1111/j.1467-8659.2003.00722.x.

Hecht, H., Bertamini, M., & Gamer, M. (2005). Naive optics: Acting on mirror reflections. *Journal of Experimental Psychology: Human Perception and Performance,* 31(5), 1023–1038.

Heider, F., & Simmel, M. (1944). An experimental study of apparent behavior. *American Journal of Psychology,* 57(2), 243. doi:10.2307/1416950.

Helmholtz, H. von (2011). *Helmholtz's treatise on physiological optics* (Vol. 1). J. P. C. Southall (Ed.). Rochester, NY: Optical Society of America.

Hering, E. (1878). *Zur Lehre vom Lichtsinne sechs Mittheilungen an die Kaiserl. Akademie der Wissenschaften in Wien* (2nd ed.). Vienna: Gerold's Sohn.

Hering, E. (1964). *Outlines of a theory of the light sense* (L. M. Hurvich & D. Jameson, Trans.). Cambridge, MA: Harvard University Press.

Hershberger, W. (1970). Attached-shadow orientation perceived as depth by chickens reared in an environment illuminated from below. *Journal of Comparative and Physiological Psychology*, 73(3), 407–411. doi:10.1037/h0030223.

Hopkins, G. (1883). Shadow-beams in the east at sunset. *Nature*, 29(733), 55. doi: 10.1038/029055b0.

Horn, B. (1975). Obtaining shape from shading. In P. H. Winston (Ed.), *The psychology of computer vision* (pp. 115–155). New York: McGraw-Hill.

Howard, I. P. (1996). Alhazen's neglected discoveries of visual phenomena. *Perception*, 25(10), 1203–1217.

Hubona, G. S., Shirah, G. W., & Jennings, D. K. (2004). The effects of cast shadows and stereopsis on performing computer-generated spatial tasks. *IEEE Transactions on Systems, Man, and Cybernetics—Part A: Systems and Humans*, 34(4), 483–493. doi: 10.1109/TSMCA.2004.826269.

Imura, T., Shirai, N., Tomonaga, M., Yamaguchi, M. K., & Yagi, A. (2008). Asymmetry in the perception of motion in depth induced by moving cast shadows. *Journal of Vision*, 8(13), 10.1–8. https://doi.org/ 10.1167/8.13.10.

Imura, T., & Tomonaga, M. (2003). Perception of depth from shading in infant chimpanzees (*Pan troglodytes*). *Animal Cognition*, 6(4), 253–258. doi:10.1007/s10071 -003-0188-5.

Jackendoff, R. (2010). *Meaning and the lexicon: The parallel architecture, 1975–2010*. Oxford: Oxford University Press.

Jacobson, J., & Werner, S. (2004). Why cast shadows are expendable: Insensitivity of human observers and the inherent ambiguity of cast shadows in pictorial art. *Perception*, 33(11), 1369–1383. doi:10.1068/p5320.

Johnson, M. K., & Farid, H. (2007a). Exposing digital forgeries through specular highlights on the eye. Presented at the 9th International Workshop on Information Hiding, Saint Malo, France.

Johnson, M., & Farid, H. (2007b). Exposing digital forgeries in complex lighting environments. *IEEE Transactions on Information Forensics and Security*, 2, 450–461.

Kahneman, D., & Tversky, A. (1996). On the reality of cognitive illusions. *Psychological Review, 103*(3), 582–591.

Kanizsa, G. (1979). *Organization in vision.* Santa Barbara, CA: Praeger.

Kawabe, T., Maruya, K., & Nishida, S. (2015). Perceptual transparency from image deformation. *Proceedings of the National Academy of Sciences of the United States of America, 112*(33), E4620–E4627. doi:10.1073/pnas.1500913112.

Kee, E., O'Brien, J. F., & Farid, H. (2013). Exposing photo manipulation with inconsistent shadows. *ACM Transactions on Graphics (ToG), 32*, 1–12. doi:10.1145/2487228.2487236.

Kellman, P. J., & Shipley, T. F. (1991). A theory of visual interpolation in object perception. *Cognitive Psychology, 23*(2), 141–221.

Kennedy, J. M. (1974). *A psychology of picture perception.* San Francisco: Jossey-Bass.

Kersten, D., Knill, D. C., Mamassian, P., & Bülthoff, I. (1996). Illusory motion from shadows. *Nature, 379*(6560), 31. doi:10.1038/379031a0.

Kersten, D., Mamassian, P., & Knill, D. C. (1991). Moving cast shadows generate illusory object trajectories. *Investigative ophthalmology and visual science, 32*, 1179.

Kersten, D., Mamassian, P., & Knill, D. (1994). *Moving cast shadows and the perception of relative depth.* Max Planck Institute for Biological Cybernetics Technical Report No. 6.

Kersten, D., Mamassian, P., & Knill, D. C. (1997). Moving cast shadows induce apparent motion in depth. *Perception, 26*(2), 171–192. doi:10.1068/p260171.

Khuu, S. K., Khambiye, S., & Phu, J. (2012). Detecting the structural form of cast shadows patterns. *Journal of Vision, 12*(11), 1–14. doi:10.1167/12.11.10.

King, D. (1997). *The commissar vanishes: The falsification of photographs and art in Stalin's Russia.* New York: Metropolitan Books.

Kingdom, F. A. A., Beauce, C., & Hunter, L. (2004). Colour vision brings clarity to shadows. *Perception, 33*(8), 907–914. doi:10.1068/p5264.

Knill, D. C., Mamassian, P., & Kersten, D. (1997). Geometry of shadows. *Journal of the Optical Society of America A: Optics, Image Science, and Vision, 14*(12), 3216–3232.

Koenderink, J. J. (1984). What does the occluding contour tell us about solid shape? *Perception, 13*(3), 321–330. doi:10.1068/p130321.

Koenderink, J., van Doorn, A., Pont, S., & Richards, W. (2008). Gestalt and phenomenal transparency. *Journal of the Optical Society of America A: Optics, Image Science, and Vision, 25*(1), 190–202.

Koenderink, J., van Doorn, A., Pont, S., & Wijntjes, M. (2010). Phenomenal transparency at X-junctions. *Perception*, 39(7), 872–883. doi:10.1068/p6528.

Koffka, K. (1935). *Principles of Gestalt psychology*. New York: Harcourt, Brace.

Kretzenbacher, L. (1961). Die Legende vom heilenden Schatten. *Fabula*, 4(2), 231–247.

Kris, E., & Kurz, O. (1934). *Die Legende vom Künstler: Ein geschichtlicher Versuch*. Vienna.

Leo, I., & Simion, F. (2009). Newborns' Mooney-face perception. *Infancy*, 14(6), 641–653. https://doi.org/10.1080/15250000903264047.

Liden, W. H., & Herberholz, J. (2008). Behavioral and neural responses of juvenile crayfish to moving shadows. *Journal of Experimental Biology*, 211(9), 1355–1361. doi:10.1242/jeb.010165.

Lindberg, D. C. (1967). Alhazen's theory of vision and its reception in the West. *Isis*, 58(3), 321–341. doi:10.1086/350266.

Lindberg, D. C. (1976). *Theories of vision from Al-Kindi to Kepler*. Chicago: University of Chicago Press.

Livingston, W., & Lynch, D. (1979). Mountain shadow phenomena. *Applied Optics*, 18(3), 265–269. doi:10.1364/AO.18.000265.

Logvinenko, A. D. (1999). Lightness induction revisited. *Perception*, 28(7), 803–816.

Logvinenko, A. D., Adelson, E. H., Ross, D. A., & Somers, D. (2005). Straightness as a cue for luminance edge interpretation. *Attention, Perception, and Psychophysics*, 67(1), 120–128.

Lopez-Moreno, J., Sundstedt, V., Sangorrin, F., & Gutierrez, D. (2010). Measuring the perception of light inconsistencies. *Proceedings of the 7th ACM Symposium on Applied Perception in Graphics and Visualization*, 7, 25–32. doi:10.1145/1836248.1836252.

Lovell, P. G., Gilchrist, I. D., Tolhurst, D. J., & Troscianko, T. (2009). Search for gross illumination discrepancies in images of natural objects. *Journal of Vision*, 9(1), 37.

Ludovici, S. S. (1954). *Miniature di Belbello da Pavia*. Milano: Aldo Martello Editore.

Mac Cumhaill, C. (2011). Specular space. *Proceedings of the Aristotelian Society*, 111 (pt. 3), 487–495. doi:10.1111/j.1467-9264.2011.00320.x.

Mamassian, P. (2004). Impossible shadows and the shadow correspondence problem. *Perception*, 33(11), 1279–1290. https://doi.org/10.1068/p5280.

Mamassian, P., & Goutcher, R. (2001). Prior knowledge on the illumination position. *Cognition*, 81, B1–B9. doi:10.1016/S0010-0277(01)00116-0.

Mamassian, P., & Kersten, D. (1996). Illumination, shading and the perception of local orientation. *Vision Research*, 36(15), 2351–2367.

Mamassian, P., Knill, D. C., & Kersten, D. (1998). The perception of cast shadows. *Trends in Cognitive Sciences*, 2(8), 288–295.

Margolis, E., & Laurence, S. (Eds.). (1999). *Concepts: Core readings.* Cambridge, MA: MIT Press.

Marr, D. (1982). *Vision: A computational investigation into the human representation and processing of visual information.* San Francisco: W. H. Freeman.

Masin, S. C. (2006). Test of models of achromatic transparency. *Perception*, 35(12), 1611–1624. doi:10.1068/p5034.

Meiss, M. (1968). Some remarkable early shadows in a rare type of Threnos. In *Festschrift Ulrich Middeldorf* (pp. 112–118). Berlin: Walter de Gruyter.

Metelli, F. (1974). The perception of transparency. *Scientific American*, 230(4), 90–98. doi:10.1038/scientificamerican0474-90.

Metelli, F. (1985). Stimulation and perception of transparency. *Psychological Research*, 47(4), 185–202.

Michotte, A. (1963). *The perception of causality.* New York: Basic Books.

Miller, J. (1998). *On reflection.* New Haven, CT: National Gallery Publications.

Minnaert, M. (2013). *Nature of light and colour in the open air.* Mineola, NY: Dover.

Mooney, C. M. (1957). Age in the development of closure ability in children. *Canadian Journal of Psychology*, 11(4), 219–226.

Moore, C., & Cavanagh, P. (1998). Recovery of 3D volume from 2-tone images of novel objects. *Cognition, 67*, 45–71.

Morgenstern, Y., Murray, R. F., & Harris, L. R. (2011). The human visual system's assumption that light comes from above is weak. *Proceedings of the National Academy of Sciences, 108*, 12551–12553. doi:10.1073/pnas.1100794108.

Mougenel, R. (2011). *La pêche miraculeuse de Konrad Witz: Visions dynamiques des peintures du retable de Genève.* Geneva: Editions Notari.

Mullen, K. T. (1991). Colour vision as a post-receptoral specialization of the central visual field. *Vision Research, 31*(1), 119–130. https://doi.org/10.1016/0042-6989(91)90079-K.

Mullen, K. T., & Kingdom, F. A. A. (1991). Colour contrast in form perception. In P. Gouras (Ed.), *The Perception of Colour* (pp. 198–217), vol. 6 of J. Cronly-Dillon (Ed.), *Vision and Visual Dysfunction.* London: Macmillan.

Nakayama, K., & Shimojo, S. (1992). Experiencing and perceiving visual surfaces. *Science*, 257, 1357–1363.

Nakayama, K., Shimojo, S., & Ramachandran, V. S. (1990). Transparency: Relation to depth, subjective contours, luminance, and neon color spreading. *Perception*, 19(4), 497–513. doi:10.1068/p190497.

Noë, A. (2003). Causation and perception: The puzzle unravelled. *Analysis*, 63(278), 93–100.

Norman, J. F., Dawson, T. E., & Raines, S. R. (2000). The perception and recognition of natural object shape from deforming and static shadows. *Perception*, 29(2), 135–148. doi:10.1068/p2994.

Norman, J. F., Lee, Y. L., Phillips, F., Norman, H. F., Jennings, L. R., & McBride, T. R. (2009). The perception of 3-D shape from shadows cast onto curved surfaces. *Acta Psychologica*, 131(1), 1–11. doi:10.1016/j.actpsy.2009.01.007.

Ollis, M., & Stentz, A. (1997). Vision-based perception for an automated harvester. *Proceedings of the 1997 IEEE/RSJ International Conference on Intelligent Robot and Systems: Innovative Robotics for Real-World Applications. IROS '97*, 3, 1838–1844. https://doi.org/ 10.1109/IROS.1997.656612.

Palmer, S. E. (1999). *Vision science: Photons to phenomenology*. Cambridge, MA: MIT Press.

Pensabene, P., & Gallocchio, E. (2011). The Villa del Casale of Piazza Armerina. *Expedition*, 53(2), 29–37.

Perdeck, A. C., & Tinbergen, N. (1951). On the stimulus situation releasing the begging response in the newly hatched herring gull chick (*Larus argentatus argentatus Pont.*). *Behaviour*, 3(1), 1–39. doi:10.1163/156853951X00197.

Peterson, M. A., & Gibson, B. S. (1991). The initial identification of figure-ground relationships: Contributions from shape recognition processes. *Bulletin of the Psychonomic Society*, 29(2), 199–202. doi:10.3758/BF03335234.

Plummer, D. J., & Ramachandran, V. S. (1993). Perception of transparency in stationary and moving images. *Spatial Vision*, 7(2), 113–123.

Poncelet, J.-V. (1822). *Traité des propriétés projectives des figures*. Paris: Bachelier. http://docnum.unistra.fr/cdm/ref/collection/coll7/id/72178.

Pont, S. C., Wijntjes, M. W. A., Oomes, A. H. J., van Doorn, A., van Nierop, O., de Ridder, H., et al. (2011). Cast shadows in wide Perspective. *Perception*, 40(8), 938–948. doi:10.1068/p6820.

Prinz, J. J. (2002). *Furnishing the mind: Concepts and their perceptual basis*. Cambridge, MA: MIT Press.

Pylyshyn, Z. (1999). Is vision continuous with cognition? The case for cognitive impenetrability of visual perception. *Behavioral and Brain Sciences*, 22(3), 341–365, discussion 366–423.

Qiu, F. T., Sugihara, T., & von der Heydt, R. (2007). Figure-ground mechanisms provide structure for selective attention. *Nature Neuroscience*, 10(11), 1492–1499. doi:10.1038/nn1989.

Quiroga, R. Q., Kreiman, G., Koch, C., & Fried, I. (2008). Sparse but not "grandmother-cell" coding in the medial temporal lobe. *Trends in Cognitive Sciences*, 12(3), 87–91.

Ramachandran, V. S., Altschuler, E. L., & Hillyer, S. (1997). Mirror agnosia. *Proceedings of the Royal Society B: Biological Sciences*, 264(1382), 645–647. doi:10.1098/rspb.1997.0091.

Rensink, R. A., & Cavanagh, P. (2004). The influence of cast shadows on visual search. *Perception*, 33(11), 1339–1358. doi:10.1068/p5322.

Rensink, R. A., & Enns, J. T. (1998). Early completion of occluded objects. *Vision Research*, 38(15), 2489–2505.

Rock, I. (1985). *The logic of perception.* Cambridge, MA: MIT Press.

Royal Collection Trust. (2017). Secrets of Canaletto's drawings revealed ahead of a new exhibition at the Queen's Gallery, Buckingham Palace. Press release, April 2017. https://d9y2r2msyxru0.cloudfront.net/sites/default/files/resources/Canaletto_Infrared%20photography.pdf.

Rubin, J. M., & Richards, W. A. (1982). Color vision and image intensities: When are changes material? *Biological Cybernetics*, 45(3), 215–226.

Sabler, L. (2004). *Dalí.* London: Haus Publishing.

Santos, P. E., Casati, R., Dee, H. M., Schultz, C., & Bhatt, M. (2015). Eclipse in occlusion: A perspectival mereotopological representation of celestial eclipses. https://ivi.fnwi.uva.nl/tcs/QRgroup/qr16/pdf/13Santos.pdf.

Sattler, M., Sarlette, R., Mücken, T., & Klein, R. (2005). Exploitation of human shadow perception for fast shadow rendering. In *Proceedings of the 2nd Symposium on Applied Perception in Graphics and Visualization—APGV '05* (pp. 131–134). La Coruña, Spain, August 26–28, 2005. https://doi.org/ 10.1145/1080402.1080426.

Sayim, B., & Cavanagh, P. (2011). The art of transparency. *i-Perception*, 2(7), 679–696. doi:10.1068/i0459aap.

Schoon, K. J. (1993). The origin of earth and space science misconceptions: A survey of preservice elementary teachers. In *Proceedings of the Third International Seminar on Misconceptions and Educational Strategies in Science and Mathematics*. Ithaca, NY: Misconceptions Trust.

Segal, G. (1989). Seeing what isn't there. *Philosophical Review*, 98, 189–214.

Sharpe, W. (2017). *Grasping shadows: The dark side of literature, painting, photography, and film*. New York: Oxford University Press.

Shearman, J. (1966). Masaccio's Pisa altar-piece: An alternative reconstruction. *Burlington Magazine*, 108(762), 449–455.

Smith, A. M. (1996). Ptolemy's theory of visual perception: An English translation of the "Optics" with introduction and commentary. *Transactions of the American Philosophical Society*, 86(2), iii–300. doi:10.2307/3231951.

Sørensen, R. A. (2007). *Seeing dark things: The philosophy of shadows*. New York: Oxford University Press.

Spelke, E. S. (1990). Principles of object perception. *Cognitive Science*, 14(1), 29–56.

Sperber, D., & Hirschfeld, L. A. (2004). The cognitive foundations of cultural stability and diversity. *Trends in Cognitive Sciences*, 8(1), 40–46.

Staddon, J. E. R. (1975). A note on the evolutionary significance of "supernormal" stimuli. *American Naturalist*, 109(969), 541–545. doi:10.1086/283025.

Stoffregen, T. A. (2013). On the physical origins of inverted optic images. *Ecological Psychology*, 25(4), 369–382. doi:10.1080/10407413.2013.839896.

Stoichita, V. I. (1997). *A short history of the shadow*. London: Reaktion Books.

Stoner, G. R. (1999). Transparency. In R. A. Wilson & F. C. Keil (Eds.), *The MIT encyclopedia of cognitive sciences* (pp. 845–847). Cambridge, MA: MIT Press.

Stoner, G. R., Albright, T. D., & Ramachandran, V. S. (1990). Transparency and coherence in human motion perception. *Nature*, 344(6262), 153–155.

Suddendorf, T., & Whiten, A. (2001). Mental evolution and development: Evidence for secondary representation in children, great apes, and other animals. *Psychological Bulletin*, 127(5), 629–650. doi:10.1037/0033-2909.127.5.629.

Sugano, N., Kato, H., & Tachibana, K. (2003). The effects of shadow representation of virtual objects in augmented reality. In *The Second IEEE and ACM International Symposium on Mixed and Augmented Reality, 2003: Proceedings* (pp. 76–83). IEEE. Retrieved from http://ieeexplore.ieee.org/abstract/document/1240690.

Sun, J., & Perona, P. (1998). Where is the sun? *Nature Neuroscience*, 1, 183.

Switkes, E., Bradley, A., & De Valois, K. K. (1988). Contrast dependence and mechanisms of masking interactions among chromatic and luminance gratings. *Journal of the Optical Society of America A*, 5, 1149–1162.

Talmy, L. (1983). How language structures space. In H. L. Pick & L. P. Acredolo (Eds.), *Spatial orientation: Theory, research, and application* (pp. 225–282). Boston, MA: Springer.

Tappen, M. F., Freeman, W. T., & Adelson, E. H. (2005). Recovering intrinsic images from a single image. *IEEE Transactions on Pattern Analysis and Machine Intelligence*, *27*(9), 1459–1472.

Tarr, M. J., Kersten, D., & Bülthoff, H. H. (1998). Why the visual recognition system might encode the effects of illumination. *Vision Research*, *38*, 2259–2275. doi:10 .1016/S0042-6989(98)00041-8.

Tarr, M. J., & Bülthoff, H. H. (1998). Image-based object recognition in man, monkey and machine. *Cognition*, *67*(1), 1–20.

Taylor, P. (2015). *Condition: The ageing of art*. London: Paul Holberton.

Todes, S. (1975). Part II. Shadows in knowledge: Plato's misunderstanding of shadows and of knowledge as shadow-free. In D. Ihde & R. M. Zaner (Eds.), *Dialogues in phenomenology* (pp. 94–113). Amsterdam: Springer Netherlands.

Todes, S., & Daniels, C. (1975). Part I. Beyond the doubt of a shadow. In D. Ihde & R. M. Zaner (Eds.), *Dialogues in phenomenology* (pp. 86–93). Amsterdam: Springer Netherlands.

Toyama. K. (2009). Light and shadow in Sassetta: *The Stigmatization of Saint Francis* and the sermons of Bernardino da Siena. In M. Israëls (Ed.), *Sassetta: The Borgo San Sepolcro altarpiece* (pp. 305–318). Cambridge, MA: Harvard University Press.

Toyama, K., & Naito, M. (2007). A comparative survey on the rendering of shade and shadow in painting: Edo period Japan and Italian early Renaissance. *CARLS Series of Advanced Study of Logic and Sensibility*, 1, 325–338.

Travi, C. (2011). *La regalita di Cristo: Pitture murali in Sant'Abbondio a Como*. Milano: Skira.

Tse, P. U. (1999a). Complete mergeability and amodal completion. *Acta Psychologica*, 102(2–3), 165–201.

Tse, P. U. (1999b). Volume completion. *Cognitive Psychology*, 39(1), 37–68. doi:10.1006/cogp.1999.0715.

Tversky, B. (2017). Shadow play. *Spatial Cognition and Computation*, *18*, 86–96. doi:1 0.1080/13875868.2017.1331442.

Ullman, S. (1996). *High-level vision: Object recognition and visual cognition*. Cambridge, MA: MIT Press.

Van de Walle, G. A., Rubenstein, J. S., & Spelke, E. S. (1998). Infant sensitivity to shadow motions. *Cognitive Development*, 13(4), 387–419. doi:10.1016/S0885-2014 (98)90001-6.

Volwahsen, A. (2001). *Cosmic architecture in India: The astronomical monuments of Maharaja Jai Singh II*. Munich: Prestel.

Walton, K. L. (1990). *Mimesis as make-believe: On the foundations of the representational arts*. Cambridge, MA: Harvard University Press.

Waltz, D. L. (1972). Generating semantic descriptions from drawings of scenes with shadows. *Technical Report*, 271, 349.

Waltz, D. L. (1975). Understanding line drawings of scenes with shadows. In P. H. Winston (Ed.), *The psychology of computer vision*. New York: McGraw-Hill.

Watanabe, T., & Cavanagh, P. (1996). Texture laciness: The texture equivalent of transparency? *Perception*, 25(3), 293–303. doi:10.1068/p250293.

Wijntjes, M. W., & de Ridder, H. (2014). Shading and shadowing on Canaletto's Piazza San Marco. In *IS&T/SPIE Electronic Imaging* (Vol. 9014, p. 901415). International Society for Optics and Photonics. https://doi.org/ 10.1117/12.2047854.

Winkler, A. D., Spillmann, L., Werner, J. S., & Webster, M. A. (2015). Asymmetries in blue-yellow color perception and in the color of "the dress." *Current Biology*, 25(13), R547–R548. doi:10.1016/j.cub.2015.05.004.

Wong, D. (2010). *The Wall Street Journal guide to information graphics*. New York: Norton.

Xia, L., Pont, S. C., & Heynderickx, I. (2016). Effects of scene content and layout on the perceived light direction in 3D spaces. *Journal of Vision*, *16*, 1–13. doi:10.1167/ 16.10.14.

Yonas, A., Goldsmith, L. T., & Hallstrom, J. L. (1978). Development of sensitivity to information provided by cast shadows in pictures. *Perception*, 7(3), 333–341.

Yonas, A., & Granrud, C. E. (2006). Infants' perception of depth from cast shadows. *Perception and Psychophysics*, 68(1), 154–160.

Zacks, J. M., & Tversky, B. (2001). Event structure in perception and conception. *Psychological Bulletin*, 127(1), 3–21.

Index